Mountains: A Very Short Introduction

'A masterful treatment of virtually all aspects of Mountains,
outstanding, well-documented, and highly readable. It should prove
invaluable to all practitioners and involved scientists and an excellent
introduction to the great majority of humankind, who are now
awakening to problems facing our World.'

Jack Ives, Professor Emeritus at Carleton University, Former coordinator
of UNU Mountain Project (1978–2002)

VERY SHORT INTRODUCTIONS are for anyone wanting a stimulating and accessible way into a new subject. They are written by experts, and have been translated into more than 40 different languages.

The series began in 1995, and now covers a wide variety of topics in every discipline. The VSI library now contains over 400 volumes—a Very Short Introduction to everything from Psychology and Philosophy of Science to American History and Relativity—and continues to grow in every subject area.

## Very Short Introductions available now:

## Available soon:

For more information visit our website

www.oup.com/vsi/

Martin F. Price

# MOUNTAINS

## A Very Short Introduction

# OXFORD
UNIVERSITY PRESS

Great Clarendon Street, Oxford, OX2 6DP,
United Kingdom

Oxford University Press is a department of the University of Oxford.
It furthers the University's objective of excellence in research, scholarship,
and education by publishing worldwide. Oxford is a registered trade mark of
Oxford University Press in the UK and in certain other countries

Published in the United States of America by Oxford University Press
198 Madison Avenue, New York, NY 10016, United States of America

British Library Cataloguing in Publication Data
Data available

Library of Congress Control Number: 2015935343

ISBN 978-0-19-969588-1

Printed in Great Britain by
Ashford Colour Press Ltd, Gosport, Hampshire

*For Randi, Katia, and Kamilla*

# Contents

# Preface

Mountains have been an important part of my life since my parents first took me to North Wales at the age of 5 and to the Swiss Alps a year later. I remember standing under the blue ice of the snout of the Morteratsch glacier. Fifty years later, I returned to the same valley and found that the glacier had retreated more than a kilometre. Trees higher than me grew where the snout of the glacier had been. The valley was still beautiful, but rather different. It very much brought back to me how many changes have occurred in our mountains over the past half century.

Until I was about 20, my main interest in mountains was to climb them. After my first degree, they became the focus of my professional life, and enjoying them remained important in my personal life—a shared love of mountains was one of the reasons that brought my wife, Randi, and me together. As a scientist, my initial interest in mountains was their environments, but I soon realized that 'environmental issues' are really about how people interact with their environments, so that many challenges associated with the mountains require inter-disciplinary perspectives. After completing my doctorate in Boulder, Colorado, my emphasis shifted into understanding the human dimensions of climate change; but I soon returned to a focus on mountains, becoming involved in the preparation of the mountain chapter of 'Agenda 21' for the 1992 Rio Earth Summit. Since then, I have been involved with many mountain initiatives, giving me the opportunity to travel to many mountain areas, see their remarkable landscapes, experience all types of weather, meet many wise and committed people, and enjoy many aspects of mountain cultures—not least the food, which always tastes better in the mountains! I have also worked with many international organizations, experiencing the challenges of trying to make progress, which is often rather gradual, even when the rhetoric is there—yet very rewarding when there is a breakthrough in cooperation or understanding. This often comes from a realization of common interests; and sharing knowledge has been an important part of my life, including convening many meetings, editing books, giving talks, and writing for different media. Moving towards an understanding of the very complex interactions of mountain environments and their people is always decidedly challenging. Yet, if scientists and policy-makers work together—and interact constructively with local people—it is possible to gain, share, and act on knowledge to the benefit of not only the millions of people who live in mountain areas, but also the billions who depend on them in some way, even if they may not realize it.

# List of illustrations

# Chapter 1
# Why do mountains matter?

## What is a mountain?

Mountains are found on every continent, and from the Equator—which is straddled by Mount Kenya and is just north of the active volcano of Pinchincha, next to Quito, Ecuador's capital—almost to both poles. The Queen Maud Mountains extend beyond 85°S in Antarctica, with their highest peak being Mount Kaplan, 4,320 metres high; and, in the Arctic, the mountainous islands of Greenland and Ellesmere Island stretch beyond 83°N. But what is a mountain? Everyone can agree that Mount Kenya, Pinchincha, and Mount Kaplan are indeed mountains—as well as better known peaks such as Mount Everest, the Matterhorn, and Mount Fuji. But how high does a hill have to be before it is to be called a mountain? The highest place in the Netherlands is the Vaalserberg, which can be translated as Vaals mountain. It is 323 metres high. In the USA, Iron Mountain, Florida, is 100 metres high. So, while every mountain has a summit, how high must this be? Altitude can certainly not be a criterion on its own.

As anyone who has walked in mountainous terrain knows, mountains go up and down. Walking up a mountain takes effort! In other words, mountains have steep slopes of some length. So, although the altitude of the South Pole is 2,800 metres, the 'mile-high city' of Denver on the Great Plains to the west of the Rockies in Colorado is at 1,600 metres, and the Tibetan plateau has an average elevation of 4,500 metres, these are all rather flat environments that no one would characterize as mountainous—although the altitude on its own might make you short of breath. They—and probably the Vaalserberg and Iron Mountain—also do not meet the criteria suggested in 1936 by the American geographer Roderick Peattie: that mountains should be 'impressive: they should enter into the imagination of the people who live within their shadows', and that they should have bulk and individuality. However, these criteria are subjective; and perceptions change over time. For example, the first explorers in the area around the Gulf of St. Lawrence clearly identified the 'lofty' Wotchish mountains, some 500 metres high. Today, they rarely appear on maps, and are recognized as just part of the vast Labrador plateau. And again, locally used terms differ: some Scots say that their country has no mountains, only hills—though few would dispute that parts of Scotland are distinctly mountainous; and the Middle Hills of Nepal extend to about 3,000 metres in height.

As Peattie noted, mountains have bulk, and this can be measured in terms of the relief of a landscape: the difference in altitude between its highest and lowest points or, in other words, the roughness of the topography. Again, if only the criterion of height difference is used, some anomalies may occur. For instance, the Grand Canyon is 1,600 metres deep but, as a canyon, it effectively has negative bulk. So, to delineate the world's mountains requires methods to measure and compare topography. Until recently, this was done using traditional surveying instruments. However, over the past two decades, satellites circling the globe have allowed us to measure altitude from space and then, using computerized mapping systems, to calculate the relief of the Earth's landforms. This means that, instead of previous debates about what is or is not a mountain, based on local or regional criteria or perceptions, it has been possible to agree global criteria.

The starting point was a database, developed by the US Geological Survey in the 1990s, that recorded the average altitude of every square kilometre of the Earth's surface. Criteria relating to altitude, slope, and relief were developed, and scientists, mountaineers, and policy-makers were asked to evaluate different combinations of these criteria for the mountain areas they knew best. It was agreed that all parts of the Earth more than 2,500 metres above sea level are mountainous because of the physiological challenges to living there, and that high-altitude plateaux and plains should be 'filtered out' by setting an appropriate slope threshold. The great innovation of this work was that it included the evaluation of relief—something that, to be done precisely and consistently, required such a detailed, global database. The criteria that were eventually chosen were that, for a place to be defined as mountainous, the altitude had to vary by at least 300 metres over a radius of 7 kilometres. The resulting map, published in 2000, showed not only the world's higher mountains, but also the older and lower mountain systems, such as the Appalachians, the Scottish Highlands, the Atlantic Highlands of Brazil, the Urals, the Australian Alps, and the mountains of southeastern China (see Figure 1).

This analysis showed that mountains cover 35.8 million square kilometres: 24 per cent of the Earth's land surface. From this basis, it was possible to undertake other analyses in order to evaluate the global importance of mountains, overlaying the map of mountains with maps based on other global databases. One of the longest running debates had been how many people lived in mountains. For some time, the estimate had been 10 per cent of the world's population. Also in the 1990s, the first global map of population had been created, using both census data and the light visible at night from space to identify urban areas. Once this was overlain with the global map of mountains in 2003, it was calculated that 720 million people, 12 per cent of the world's population, lived in mountains; and a further 14 per cent lived very close to them. Thus, since the early years of this century, it

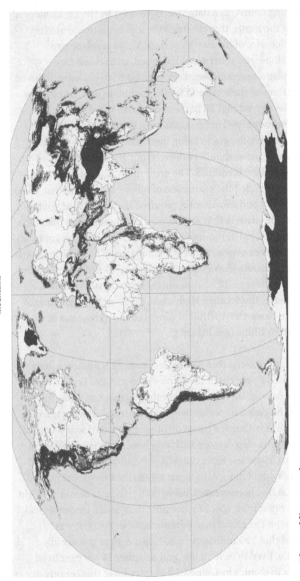

1. The world's mountains.

has been clear that mountains matter in quantitative terms, covering a quarter of the Earth's land surface, with about a quarter of humanity living in and around them.

## Historical perspectives

There are three main reasons why mountains have been important for people at very broad scales, from regional to intercontinental, for centuries if not millennia: as sources of food and of valuable minerals and precious stones, and as places of great cultural importance. The first relates to one of our most fundamental needs. Eight parts of the world have been identified as being the original centres for the domestication of plants, and all are mainly or partly mountainous: along the Andes, around the Mediterranean, and in central America, Ethiopia, the Middle East, central Southeast Asia, China, and India. Of the twenty plants that supply 80 per cent of humanity's food, six—maize, potatoes, barley, sorghum, apples, and tomatoes—originated in mountain areas. Another seven—wheat, rice, beans, oats, grapes, oranges, and rye—were introduced into the mountains and have been transformed into many different varieties. For those of us who enjoy coffee or tea, the former originated in the mountains of Ethiopia and the latter probably in the mountains where the frontiers of China and Myanmar are now located. Similarly, many spices originate in mountain areas, including saffron, from the mountains of western Asia, possibly Iran; and black pepper and cardamom, both from India's Western Ghats, the basis of an international trading economy as early as 3000 BC. Many domestic animals were certainly or probably first domesticated in mountain areas, such as alpaca, goats, llama, sheep, and yaks. Farmers still actively maintain some of the great centres where this great agrobiodiversity emerged, as around Cusco in Peru, with more wild potato species than anywhere else and more than 400 of the currently used potato varieties; some farmers grow up to a hundred in one plot to minimize risks and maximize the likely yields.

Potatoes are a prime example of the value and, in some cases, risks of the global spread of crops from mountain areas. When they were introduced to the Swiss Alps and the Khumbu of Nepal, for instance, farmers realized that potatoes yielded far more than the traditional cereals—and populations grew rapidly in the following years. The introduction of potatoes had similar consequences in many lowland areas. Yet, as millions found to their cost during the European potato famines of the 1840s and 1850s, too great a dependence on a narrow range of varieties of one crop can be risky. Hence, one of the values of the maintenance of original varieties in the mountains is the possibility to breed disease-resistant strains. While potatoes have become globally important, many of the other crops that have been grown around Cusco and elsewhere in the Andes for centuries are also widely adaptable, highly nutritious, and tasty. Yet only one of these has started to be grown and known widely outside the Andes: quinoa, now found in many supermarkets, as well as health food shops. Oca was introduced to Europe in the 1830s, but never grown widely there, in contrast to New Zealand, where it was introduced in 1860 and is known as yam. Compared to potatoes, it has similar nutritional qualities, can grow under harsher conditions, and can yield twice as much. And there are many others. Tarwi contains as much as, or more, protein than the world's main protein crops—beans, peas, peanuts, and soybeans—and up to 20 per cent oil, as much as soybeans. It grows in marginal soils and, as it is a lupin, adds nitrogen to the soil. Kiwicha and other amaranths are highly nutritious; their proteins are far more suitable for people than most cereals. In a world with a growing population and changing environments, mountain crops may become ever more important.

The crops that originate in mountain areas are the product of evolution and human ingenuity. The minerals found in these areas have a far longer history: the extreme temperatures and pressures that created the mountains also helped to create and concentrate minerals that people could use. Most of the world's major sources of metals are related to mountain-building, although some of the mountains have since been eroded away, and the minerals they contained have been distributed, mainly by rivers—some of which also no longer exist. So, while the metals originated in mountains, many deposits may now be far from them. One of the world's oldest known mining sites is in Wadi Faynan at the edge of the mountains of Edom in present-day Jordan, where copper has been mined for over 8,000 years, and there has been significant pollution from smelting since the Early Bronze Age, about 5,000 years ago. Somewhat later, this copper was used to produce coins that were used around the Mediterranean in the Roman and Byzantine eras. The exploitation and trading of metals from mountain areas has been an essential element in the growth and expansion of all imperial powers, starting with the Romans. They mined iron in the mountains of Austria and Romania; gold near Aosta in the Italian Alps and in eastern Serbia; and copper in the French Alps and the Sierra Morena in Spain. These are just some examples; not all Roman mining took place in mountain areas, and they exploited and traded many more metals from other places, some as far away as India and Ethiopia.

From the early 16th century to 1700, the Spanish Habsburg Empire was the world's greatest power. A significant factor was its mines in South America, including those at Potosí in Bolivia, a town founded in 1545 at an altitude of 4,000 metres at the base of the Cerro Rico, or rich mountain, which is largely composed of ore with a high proportion of silver—up to 25 per cent in some veins—as well as other metals (see Figure 2). By the end of the 16th century, Potosí had become the largest settlement in the Americas, with a population of at least 150,000—comparable to large European cities such as Amsterdam and London. The majority of the miners were local people; slaves were also imported. By 1824, up to 40,000 tons of silver had been produced from Cerro Rico. Subsequently, a further 10,000 tons have been produced, as well as lead, tin, and zinc. More than five centuries of mining, from pre-Columbian times onwards, have led to high levels of pollution in the water draining the mines, which flows

**2. Cerro Rico, rising above Potosi, Bolivia, the world's highest city.**

down the Rio Pilcomayo and into Argentina. This is one of many examples of metal mining and processing, albeit on a particularly large scale, where the benefits have largely left the mountains, leaving a legacy of air and water pollution that have been inevitable because most metal ores contain sulphur and other toxic elements. Particularly from the last century, as environmental concerns have grown in countries around the world, various engineering measures have been developed to minimize the risks of water pollution, such as tailings reservoirs in which toxic chemicals can settle out and be removed or treated. However, the quantities of water and solids are often very large, and such measures are not infallible, especially when risks are difficult to assess because of inadequate data.

When risks combine, and appropriate measures are not taken, the consequences for both people and the environment can be very severe. In recent years, one of the most serious examples has been the open-pit Ok Tedi mine in a remote area of Papua New Guinea, where mining of the copper and gold ore at the summit of Mount Fubilan—originally 2,095 metres high, but now more than 700 metres lower—started in 1984, when a new town was built to house the miners. Due to the lack of an effective tailings reservoir and waste retention facility, about two billion tons of untreated waste, including cyanide used for gold extraction, were discharged into the Fly River system, affecting fisheries and agriculture 1,000 kilometres downstream. It is estimated that this disaster affected 50,000 people in downstream villages, and that up to 3,000 square kilometres of forest may eventually be lost. Local communities won damages of US$28.6 million from the mining company in a court case, and one-third of the profits from the mine now go to the communities, but the mine, taken over by the Papua New Guinea government in 2013, is still open and some pollution does continue.

Thus, today, mountains remain major sources of many metals: for instance, half of the world's tungsten comes from the mountains of southern China; almost half of the world's silver from the mountains of western North America; and over a third of the global supply of copper from the Chilean Andes. However, metals are not the only products of mining in mountain areas that have been important for trade. In the Stone Age, from the sixth millennium BC, obsidian mined on Monte Arci in Sardinia was exported around the Mediterranean to make sharp weapons and tools. Many of the world's gemstones come from mountains, again because of the forces that created them; and these lightweight but valuable commodities are traded around the world. In most locations, the first mining is in alluvial deposits; underground mining takes place when these have been exhausted. For example, an early emerald mine was established in Wadi Sikait in the mountains of Egypt's Eastern Desert in the Ptolemaic period; production continued at least through Roman and Byzantine times. Today, the world's main source of emeralds is in the eastern Andes of Colombia. In Myanmar, the Mogok valley in the Shan Highlands was the world's principal source of rubies for centuries; these and other gemstones, such as sapphires, are still the basis of the local economy, in an area controlled by military authorities. Among semi-precious stones, lapis lazuli from the mines of Sar-e-Sang in the remote Kokcha valley of Afghanistan's Hindu Kush has been widely exported for six millennia. As with metals, the mining and processing of gemstones provide clear economic benefits to some people, companies, and governments. However, miners and other local people often experience significant negative impacts in terms of the loss of agricultural and forest land leading to land-use conflicts, health risks from mining and polluted water, and security risks where the deposits are controlled by criminal groups. Nevertheless, as the geology of gemstone deposits usually does not create toxic chemicals, very large-scale pollution is less likely than with metal mining.

A third long-term global perspective on mountains is as places with great cultural importance. To some extent, this may be contrasted with the perspective of mountains as the source of exploitable resources, for instance through mining. It is notable that, of the world's early imperial powers, the Romans generally had a negative view of mountains, as wastelands and obstacles to expansion and trade. Christians during the Medieval period and until the 18th and 19th centuries regarded mountains as dangerous, inhabited by demonic beings. Other major cultures had very different views of mountains. For instance, the Chinese regarded their mountain ranges as being the body of a cosmic being, possibly a dragon. In Chinese culture, dragons are viewed as benevolent and wise, in contrast to the malevolent dragons of Judeochristianity. The Chinese, as well as many other cultures— including Buddhists, Greeks, Hindus, Shintos, and pre-Columbian peoples in South and North America—regarded mountains as places where gods dwelled, controlling the weather and water supplies. Throughout history, people have made offerings to mountain gods to ensure that they continued to provide these blessings. They also form the focus of pilgrimages, most notably Mount Kailash, in Tibet, perhaps the most holy mountain on Earth as it is sacred to the adherents of the Hindu, Buddhist, Jain, and Bon religions—about a quarter of the world's population (see Figure 3). For Hindus, it is the home of Lord Shiva; for Buddhists,

**3. Buddhist stupas, or chorten, on the trail around Mount Kailash, Tibet.**

the deity Demchog. One reason for its importance is undoubtedly that it is near the source of four of Asia's major rivers: the Brahmaputra, the Indus and its tributary the Sutlej, and the Karnali, which flows into the Ganges.

Mountains have therefore been regarded as places of power by significant numbers of people for a very long time. While the Greeks and Incas climbed mountains to make offerings and build shrines, people from other cultures did not climb the mountains where the gods lived, because this would be sacrilegious; the summit of Kailash remains untrodden to this day. In China, this perception changed in the 4th century AD, when people began to appreciate the beauty of mountain landscapes. The term for landscape in Chinese means 'mountains and water', and mountains are often painted as emerging from clouds, sometimes shown as dragons. Until the 9th century, emperors climbed the sacred mountain of T'ai Shan to conduct sacrifices, or appealed to it during times of drought, earthquake, or flood. To the east, the Shugendo religion emerged in Japan in the 7th century, with elements of Buddhism, Shintoism, and Taoism: its followers climb mountains to gain spiritual and physical power. Major peaks were climbed from the 9th century, and those who had done this were treated with reverence. Climbing mountains remains an important element of the life of many Japanese people.

Comparable attitudes to mountains did not emerge widely in European civilization until the 18th century, when Jean-Jacques Rousseau wrote about the beautiful landscapes and pure air of the Alps, influencing others, who were encouraged to climb the peaks for aesthetic reasons; in the spirit of adventure; and, often, to make scientific measurements. In England, viewpoints were first erected in 1790 in the Lake District, for visitors to enjoy the best views, as described in guidebooks. In both Europe and North America, mountains became a favourite subject for landscape painters, widening the public appeal of the mountains. The initial rise of tourism from the mid-19th century was linked particularly to the building of railways, first in Europe, and then in India and North America, allowing increasing numbers of people to visit mountain landscapes. Mountaineering also expanded globally, with ascents of all the major peaks in the Alps in the second half of the 19th century and the establishment of Alpine Clubs in many countries. Today, mountains are primary destinations for hundreds of millions of tourists each year, travelling both within their own countries and around the world to see mountain landscapes; enjoy different cultures; and, in some cases, to enjoy physical challenges. For many, they are powerful and wild places to escape to: the counterpoint to an urbanized existence—a viewpoint espoused centuries ago in eastern Asia, but much more recently in many other parts of the world.

## Recognizing the global importance of mountains

So far, this chapter has argued that mountains matter in quantitative terms, with a significant proportion of our planet's land surface and population, and have long been globally important as sources of food, valuable minerals, and precious stones, and as places of great cultural importance. In recent decades, recognition of the global importance of mountains has increased in international debates and policy arenas; and media attention and public understanding of mountains has also changed, from a focus on mountaineering exploits and natural disasters to a wider understanding of mountain people and the environments on which they depend, particularly in relation to climate change. Some key themes are explored in Chapters 3, 5, and 6, particularly their importance as 'water towers' and as centres of not only agricultural but also natural biodiversity, with many protected areas and major tourist destinations.

The religious significance given to mountains in many parts of the world is an implicit recognition of their importance as modifiers of climates at all scales—another reason why mountains are important. Mountain regions include some of the wettest and driest places on Earth—and influence climates at all scales. Because the Himalaya are where they are, and are so high, monsoon clouds cannot cross them, and the highest rainfalls in the world are found to their south in the Indian state of Meghalaya—where the 'wettest place on Earth' may be the town of Cherrapunji or the village of Mawsynram. Both receive an average of about 11.8 metres of rainfall a year: Cherrapunji has the undisputed record for the greatest total in a year—23 metres from August 1860 to July 1861, including the most in one month: 9.3 metres in July 1861. On the other side of the Himalaya, Tibet has a dry climate. Similarly, the Andes block rain clouds coming from the Pacific Ocean, leading to an arid climate in western Argentina; for the same reason, the Great Basin and Mojave deserts are in the lee of California's Sierra Nevada.

These are all examples of the linkages between mountain ranges and wider regions. Such linkages are not only climatic and hydrological, based on the physical characteristics of mountains, but also economic and political. They involve both people living in and around the mountains and often those living far away—whether they benefit from mountain water or other resources; spend their free time in the mountains; or value mountains for religious reasons. However, in most countries, including those with the largest economies and greatest political power, the majority of people live in the lowlands, where their capitals and other major cities are found. As a result, politicians and policy-makers living in these cities have typically regarded mountain areas and their people as marginal. Conversely, only a few countries have most of their population, and their major cities, in the mountains; and these countries generally do not have great economic or political significance at the global—and often even regional—scale. Yet, over the past four decades, attention to mountain issues has grown—from almost nothing to the point where they are included in many major global events and processes.

At the first global meeting on the environment, held in Stockholm in 1972, mountain issues were not on the agenda. However, during the 1970s, scientists involved in major global research programmes, supported by UNESCO—the Man and the Biosphere (MAB) programme—and the United Nations University (UNU), initiated collaborative projects in many mountain areas around the world. In 1986, UNESCO, the United Nations Environment Programme (UNEP), and the World Climate Research Programme jointly established the first global structure to monitor changes in mountain environments—the World Glacier Monitoring Service. Thus, by the beginning of the 1990s, scientists around the world had begun to compile significant evidence of the diverse benefits provided by mountain people and their environments, as well as the many challenges facing them. This evidence, as well as the recognition that much more work was needed to address both knowledge gaps and development challenges, provided the foundation for a small group of scientists and development experts, who had worked in mountain regions around the world, to make sure that mountains would be included specifically in the agenda of the 1992 Earth Summit, held in Rio de Janeiro. This group, which called itself 'Mountain Agenda', recognized that this meeting would provide a unique opportunity to alert the global community to the diverse values of mountains and the need for concerted action. Working with representatives of the government of Switzerland, as well as other delegations from mountain countries at this UN meeting, they succeeded in ensuring that 'Agenda 21', the plan for action for the 21st century that came out of Rio, included a chapter on sustainable mountain development. This effectively put mountains on the global agenda, along with the other 'big issues' such as climate change, tropical deforestation, and desertification.

Over the next few years, governments in most parts of the world came together to consider the ways in which mountains were important to their countries. Some governments created mountain institutions or passed laws to benefit mountain people. New networks of mountain people and scientists emerged. In 1995, the Intergovernmental Panel on Climate Change included a chapter on the impacts of climate change on mountain regions in its second series of reports. Led by the government of the Central Asian state of Kyrgyzstan, momentum built up to declare an International Year of Mountains (IYM). In 1998, with the support of delegations from 130 countries, the largest number ever to support the declaration of an International Year, the United Nations General Assembly declared that this would be in 2002, providing a unique opportunity to raise awareness of the diverse values of mountains to everyone: as the slogan for the IYM stated, 'We are all mountain people'. A total of seventy-eight countries around the world established national committees to plan and undertake a vast number and range of scientific, cultural, sporting, and many other types of activities during the IYM. International meetings also took place, culminating in the Bishkek Global Mountain Summit in Bishkek, the capital of Kyrgyzstan. This produced a final document that became the basis for a UN General Assembly resolution which designated 11 December as International Mountain Day, and encourages the international community to organize on this day events at all levels to highlight the importance of sustainable mountain development.

The year 2002 also saw the year of the World Summit on Sustainable Development, held in Johannesburg. One of the global partnerships launched at this event was the Mountain Partnership, aiming to improve the well-being, livelihoods, and opportunities of mountain people, and to protect and provide stewardship of mountain environments around the world. More than a decade later, its members—governments, intergovernmental organizations, and a great variety of other organizations—continue to work towards these goals. To some extent this has been in the context of two major global processes which were influenced by the IYM: the Convention on Biological Diversity (CBD) and the Millennium Ecosystem Assessment (MEA). In 2004, the CBD published a programme of work which encourages governments and others to work towards the conservation, sustainable use, and sharing of the benefits of mountain biodiversity. In 2005, building on the considerable volume of knowledge developed since 1992, one of the chapters in the MEA focused on mountain ecosystems. This documents the manifold goods and services mountains provide to a large proportion of the world's population. It underlines the continuing challenges of sustainable development for the hundreds of millions of people living in mountain areas, many due to the fact that, although the knowledge base has grown, many gaps still remain.

During the 2000s, the UN General Assembly produced a number of reports and passed resolutions in support of sustainable mountain development; and in 2012, following a global assessment of progress over the past twenty years, the final document of the global Rio+20 meeting, 'The Future We Want', reiterated that the benefits derived from mountain regions are essential for sustainable development. However, this document also noted many challenges: that many mountain people remain marginalized and in poverty; and that mountain ecosystems are particularly vulnerable to the negative impacts of climate change. Finally, it called for governments to work with all those concerned with the future of mountain environments and the people who depend on them, and to adopt a long-term vision and holistic approaches. Mountains, and mountain people, are now firmly part of the global agenda.

# Chapter 2
# Mountains are not eternal

At the timescale of our human lives, almost all mountains seem to be eternal. While volcanic eruptions have created numerous small islands over the past century, it is rare for a new volcano to emerge on land, or for part of a mountain to disappear as a result of volcanic activity. Only one relatively small volcano has emerged during recent history: Paricutin, which started growing in a farmer's field in the Mexican state of Michoacán in February 1943. The eruption finished in 1952, by which time Paricutin had reached a height of 424 metres above the surrounding landscape. It grew relatively slowly, and its main effects were limited to covering about 25 square kilometres of farmland in lava. However, much more rapid changes in volcanoes can have significant effects. The eruption of Mount Saint Helens, in the Cascade Range of the USA in May 1980, led to the death of fifty-seven people and destroyed 600 square kilometres of forests and 300 kilometres of highways. A far larger eruption was that of Mount Pinatubo, in the Philippines, in June 1991, ejecting 10 cubic kilometres of magma and 20 million tonnes of sulphur dioxide, which spread around the globe and reduced the amount of sunlight reaching the Earth, decreasing global temperatures by 0.7°C. The global impacts of the explosion of the small Indonesian volcano of Krakatoa, in 1883, were even more significant, influencing weather patterns for five years. Yet even relatively small volcanic eruptions can have serious international effects, such as that of

Iceland's Eyjafjallajökull in April 2010, which caused the greatest level of disruption to European and trans-Atlantic air travel since World War II.

At completely different timescales, mountains come and go—and move around the planet. The low mountains of the Appalachians may once have been as high as today's highest mountain range, the Hindu Kush-Himalaya. And, since they were first formed, the Appalachians have moved north from the Equator. Mountains are directly linked to the process of plate tectonics, in which the six large and many smaller plates that constitute the Earth's surface move around the planet, and both under and over each other. The initial evidence for plate tectonics came in the late 1950s from the Earth's longest mountain range, the mid-Ocean Ridge that runs for 65,000 kilometres around the globe, mainly underwater. One of its best-known components is the mid-Atlantic Ridge, which runs for 15,000 kilometres parallel to the coastlines of Europe, Africa, and the Americas. Part of it rises 4,000 metres above the ocean floor, forming the mountains of Iceland and other islands such as the Azores and St. Helena. The dating of rocks across the ridge shows that the youngest rocks are in the centre, with the age increasing outwards from the ridge in both directions: a vital research finding that provided the key evidence supporting the theory of plate tectonics.

While volcanoes comprise an important proportion of the world's mountains—a proportion that is higher if one considers the mountains of the mid-Ocean Ridge—most mountains that rise above sea level have been formed at the margins of the Earth's tectonic plates. As they move slowly together, the sediments above them are compressed and uplifted, leading to folding and faulting. There have been three major mountain-building periods, or *orogenies*. The Caledonian orogeny took place about 550–370 million years ago, resulting in the mountains whose last remnants are now found in Scotland, Scandinavia, Greenland, and the northern Appalachians. The Hercynian, Appalachian, or

Variscan orogeny, about 200–400 million years ago, was even more extensive. During this period, North Africa collided with Europe, producing what are now the low or 'middle' mountains of the Massif Central in France, the Vosges on the French/German border, the Black Forest in Germany, and the Giant Mountains along the Czech/Polish border. Over the same period, continental collisions produced the southern Appalachians. The mountains that remain today from these two earlier orogenies are the hard cores of the original mountain chains; the sediments that formed their highest peaks were eroded long ago. The most recent orogeny is the Alpine orogeny, which started about sixty-five million years ago; the mountains that began to be created then are still growing. These include the Alps, Hindu Kush-Himalaya, and Andes, the mountains of western North America, and other ranges around the Pacific Rim. The fastest growing are the Himalaya. Satellite measurements show that they are rising at up to 10 millimetres a year at Nanga Parbat; the Andes—which are the longest terrestrial mountain range, 7,000 kilometres long—are growing at only 1–2 millimetres a year. A further group of mountains are often referred to as 'Great Escarpments'. These have been formed along the trailing edges, or passive margins, of tectonic plates. Most are hundreds to thousands of kilometres long and at least a kilometre high; they run roughly parallel to a coast and separate a high inland plateau from coastal lowlands. They include the Drakensberg of southern Africa, the Serro do Mar of Brazil, the Snowy Mountains of Australia, the Transantarctic Mountains, and the Western Ghats of India.

At present, Mount Everest—also known by local people as Sagarmatha (Nepali) and Chomolongma (Tibetan)—is usually described as the highest peak on Earth: the altitude of its summit above sea level is 8,848 metres. When its height was first confirmed as part of the Great Trigonometrical Survey of India in 1856, concluding a period during which the imperial powers competed to identify the world's highest mountain on their territory, its altitude was 8,840 metres. However, by other measures, Mauna Loa, in Hawaii, is the tallest mountain on Earth: the difference in altitude from its base on the sea floor to its summit is 10.2 kilometres, of which only 4,169 metres is above sea level. Together with its neighbour, Mauna Kea, this shield volcano is also the largest mountain on Earth, with a base about 225 kilometres across (see Figure 4). Such measures are also useful for comparing the Earth's mountains to those of other planets, which do not have oceans and therefore a sea level. By all measures, the tallest mountain identified so far on any planet is another shield volcano, Olympus Mons on Mars, about 600 kilometres wide at its base. As Mars has no plate tectonics, the location of Olympus Mons is fixed and it has continued to grow as a result of thousands of volcanic eruptions. Its altitude from base to top is nearly 22 kilometres, similar to the central peak of the Rheasilva crater on the asteroid and protoplanet Vesta. There are also other mountains more than 10 kilometres high on Mars, Saturn's moon Iapetus, and Jupiter's moon Io.

4. The volcanoes of Mauna Loa and Mauna Kea, Hawaii, c.1880.

## 'Every mountain shall be laid low' (Isaiah 40)

The volcanism, folding, and faulting that create mountains are only the first part of the story of how they are formed and sculpted. As soon as they begin to grow, a variety of processes, working together with gravity, begin to wear down the rocks of which they are made. These processes operate at many scales in both space and time. The smallest scale is that of weathering: the alteration of rock into finer particles. Physical weathering involves mechanical forces that break rocks apart or make their surfaces disintegrate. One typical characteristic of mountain areas is major changes in temperature over each twenty-four-hour period. When the freezing point is crossed, water within and between rocks freezes and cracks them in the process of frost weathering. This is probably the major physical weathering process in mountain areas; although rock surfaces also expand and contract as the sun warms them and they cool, it is the presence of water that is crucial.

While physical weathering just breaks rocks into smaller pieces, chemical weathering changes the original minerals into different ones, rounding the edges of rocks. Such chemical processes depend on temperature and moisture, and are therefore most important in the humid tropics and near sea coasts, though they also occur in colder and more arid conditions. Given the different chemical compositions and structures of different rock types, rates of weathering vary: for instance, sedimentary rocks, such as sandstones, tend to have higher rates of weathering than granite. In areas with massive limestones, chemical weathering often leads to characteristic *karst* topography. Finally, rocks may also be subject to biological weathering by fungi, algae, and bacteria. Although all of these processes operate at very small scales, their combined result, over years, centuries, and millennia, is to wear mountains down.

At a very different timescale, glaciers have been a major force of change in many mountain areas. For the last century or so, most of the world's glaciers have been melting as a result of climate change. However, during the Pleistocene Epoch, 2.6 million to 11,000 years ago, many of the world's mountains were far more glaciated than at present; and those that were not experienced far cooler climates. From above, glaciers are solid ice, imperceptibly moving downhill under the influence of gravity. However, because of the great weight of the ice above, the ice of the base of a glacier is plastic, filling every nook and cranny of the rock below. The movement of the glacier along this interface erodes the rock through three processes. Abrasion occurs as rocks embedded in the ice scratch and gouge the rock below. When the glacier has gone, the resulting scratches, or striations, show the direction in which the ice was moving. Ice containing no rocks, or very few, or containing rocks that are softer than the rock below, will mainly polish its surface. The second process is the crushing of rock by the weight of the ice above. The third, and probably most powerful, is plucking or quarrying: the process by which the glacier lifts the rock and rubble below it, including the products of crushing and of frost weathering in front of the glacier. Plucking also occurs when ice under pressure melts upstream of obstacles and refreezes downstream, so that the upstream side is smoothed, and the downstream side is steep where quarrying has occurred. When the glacier melts, the resulting *roches moutonnées* (sheep-like rocks) also show the direction in which it was moving.

These processes lead to two types of typical features. The first set comprises those resulting from the carving of the landscape: bowl-shaped cirques (or *corries*), angular and often pyramidal peaks, sharp ridges (or *arêtes*), and U-shaped and hanging valleys—which are then maintained by weathering processes. Deep U-shaped valleys may be entirely on land, such as those of Yosemite in California and Lauterbrunnen in Switzerland, or partly underwater, such as the fjords along the west coasts of Alaska, British Columbia, New Zealand, and Norway. The fast-moving glacier of Norway's Sogneford gouged away the rock more than 600 metres below the current sea level. The second set of features is formed by the debris left behind after glaciers melt—known as *till*—which can be of any size from silt to huge boulders, carried on the glacier's surface and left behind as *erratics* when the glacier melts. Till is often found as *moraines*: long ridges of loose rock that show the past limits of glaciers, sometimes blocking valleys and forming lakes. Other debris is deposited by the meltwater from the glaciers to form sinuous *eskers*, showing the path of the rivers that flowed under the ice, and fans in flatter areas at the end of glaciers or downstream. Usually, a large proportion of the till flows downstream, in rivers that have a characteristic milky colour because of their high silt content. The silt particles are highly reflective, creating the spectacular blues and greens of lakes downstream from active glaciers.

Glaciers are clearly powerful agents of the erosion of mountains. However, the large amounts of water present—at least some of the time—in most mountain areas are important in removing the products of weathering and glacial erosion, and streams and rivers also directly remove the sediments and rock across which they flow through fluvial erosion. Rates of fluvial erosion vary considerably over time: often, a very large proportion of material is removed in a rather small number of floods, which occur after major storms or at the beginning of spring, when the winter's snow melts rapidly, especially in the afternoon and early evening. The melting of glaciers also contributes to increased runoff and erosion. Fluvial erosion and deposition lead to many of the typical features of mountain landscapes, such as narrow V-shaped valleys, waterfalls, terraces, braided channels, alluvial fans, and wide flood plains. Over long timescales—thousands to millions of years—rates of glacial and fluvial erosion are similar, though estimating these rates is complicated by tectonic uplift and the fact that measurements have been over relatively short periods and small areas. The highest rates of erosion, sometimes exceeding 1 centimetre a year, are in river basins that are tectonically active, such as the Himalaya and Taiwan; or where volcanoes have recently erupted; or from temperate glaciers, such as those of Patagonia and Alaska.

In addition to glacial and fluvial erosion, a third type should be mentioned: wind erosion. Mountains, especially their higher parts, are often very windy places; sand and even small rocks can be blown around, to the discomfort of mountaineers. On planets without moisture, such as Mars and Venus, wind is the main agent of erosion. On the mountains of Earth, wind erosion is far less important; probably the most important role of wind is in moving snow around. In the context of weathering and erosion, this is important, because ridges and exposed slopes are often kept snow-free, while sheltered slopes are protected; and blowing snow can even create and maintain perennial snow patches and small glaciers in cirques.

## Hazardous landscapes

In addition to the gradual processes of weathering and erosion, parts of the mountain landscape are modified in rapid events, often known as 'natural hazards', such as mudflows, rockfalls, avalanches, and landslides. These are all natural processes, although people sometimes play a role in initiating them; they only become hazardous when people are involved. Mudflows are unpredictable, occurring when there is enough rain or snowmelt to liquefy mud so that it moves downhill, carrying rocks with it. Small rockfalls are common in mountains, and can be of particular concern when they damage roads, railways, or mountaineers, but larger ones are rare. Avalanches are the most common of these phenomena, typically reoccurring in particular locations because of specific interactions of the terrain and weather patterns. Loose snow avalanches, which occur mainly in newly fallen snow on steep slopes, are the most common type of avalanche. Slab avalanches have greater impacts on people, infrastructure, and ecosystems, and are the most common cause of death for people skiing outside ski areas. The most destructive are powder snow avalanches, which involve a cloud of dry powder snow over a dense avalanche. These can have a mass over 10 million tonnes, and move at speeds of over 300 kilometres per hour, not just down the initial avalanche path, but for several kilometres along flat valleys and even uphill for short distances. In historic times, the greatest number of people dying in avalanches was during the World War I, when 40,000–80,000 soldiers died on the Austrian-Italian front in the Alps. Nowadays, as prediction and mitigation measures improve, fewer people are killed, and less infrastructure is damaged. Yet, as more and more people live in, enjoy, and travel through the mountains in winter—not least with more people skiing, snowboarding, and snowmobiling, and new ski resorts being built in countries such as China and India—the risks are still substantial.

Despite their great destructive force, the impacts of individual avalanches on mountain environments are rarely very long-lasting, although avalanche paths can often be easily recognized because their vegetation is shorter and more flexible than in adjacent forests. In contrast, the largest landslides can leave major lasting marks on the landscape. They can move at similar speeds to those of powder snow avalanches, and also along valleys and uphill. The largest known landslide occurred about 10,000 years ago in the Zagros mountains of Iran: 50 billion tonnes of rock broke free from a mountain, descended 1,500 metres, travelled 20 kilometres, and covered an area of 274 square kilometres with a layer of rocks over 100 metres thick. It was triggered by an earthquake, but the rock may have been undercut by a river. Landslides can also be triggered by other causes, including most often rainfall, but also volcanic activity and some human activities. Thus, it is when different types of natural hazards combine that their effects are most devastating. The most extreme example in recent history took place on 31 May 1970, when an earthquake resulted in the collapse of the ice and rock near the summit of Huascaran in Peru. The subsequent landslide, moving at 480 kilometres an hour, devastated the town of Yungay, 12 kilometres away, killing 18,000 people in the town, causing 52,000 further deaths, and leaving 200,000 people homeless.

People who have long experience of a mountain area understand the locations and likelihoods of natural hazards of moderate magnitude and frequency, and they avoid dangerous areas or take measures to control hazards. Increased scientific understanding of the factors that are most likely to lead to these events also means that their likelihood of occurrence can be mapped and used for land use planning. However, such knowledge is not always used or recognized, especially when many new people move into an area, or money is to be made from building tourist facilities. Equally, when a particularly large event occurs, it may cause damage outside the 'safe zone', as in the small village of Galtür, in the Austrian Alps, where thirty-one people died after the rapid fall of 4 metres of snow led to a particularly large avalanche in February 1999. In other parts of the world, such events may have even greater effects. For instance, flash floods and landslides resulting from very heavy rain in the Indian Himalaya in June 2013 killed hundreds of people at the pilgrimage site of Kedarnath, at 3,546 metres, and about 100,000 people had to be airlifted to safety after the 17 kilometre-long access trail had been destroyed. Across the state of Uttarakhand, 4,200 villages were affected, and 3,758 lost power. Thousands of families lost their basic livelihood resources. The total loss to the Indian economy was estimated as $1.9 billion, and the economy of Uttarakhand, the worst-affected state, will require years to recover. Both of these events may be regarded as unexpected; but such events may become more common as climate change increases the number and intensity of extreme storms, and of rain- and snowfall events.

In addition to these weather-related hazards, others occur around the Earth's 1,500 active volcanoes. People have lived on and around these for thousands of years, particularly because of the fertile soils that have developed on weathered volcanic rocks—and because most erupt quite infrequently. Currently, about 500 million people live near volcanoes, including 100 million in cities including Auckland, Manila, Mexico City, Naples, Quito, and Seattle. Some of these, such as Kagoshika, a city of 680,000 people in Japan, are near persistently active volcanoes—in this case, Sakurajima, which has been erupting almost continuously since 1955 (see Figure 5). This situation is possible as the activity is not of high magnitude, and special precautions have been taken: regular evacuation drills and the building of shelters where people can take refuge from falling volcanic debris. Generally, volcanic risks are rare, but when they do occur, they can have major impacts—not only on people living near them, but regionally and even globally, as mentioned earlier. As with other hazards, scientific understanding of volcanoes is increasing, but each behaves differently and has its own set of hazards. These can be mapped to indicate the likely location of future lava flows and ash fall. In combination with monitoring—of seismic and magnetic activity, emissions of gases, ground deformation, and changes in local streams and lakes—this allows some confidence in prediction, usually weeks to months before an eruption. Consequently, people living near volcanoes have a better chance of survival—although there are always risks.

## Mining the mountains

In addition to the many natural processes, sometimes inadvertently initiated by people, that alter mountains, people have long intentionally been modifying mountains to extract minerals and rocks and other deposits left by glaciers and rivers. Clearly, these impacts only occur in relatively restricted zones, but the resulting changes can be significant. The focus here is on the removal of rock and other material, though it should be realized that this often follows the intentional removal of forests and other vegetation and, similarly, can have significant impacts on

5. Kagoshima city, Japan, with nearby Sakurajima volcano erupting, on 30 March 2010.

terrestrial, fluvial, and marine ecosystems—often far downstream or downwind—through various types of pollution, as discussed in Chapter 1.

The Romans were the first to develop large-scale mining methods, initially using hydraulic mining, in which water was transported by aqueducts and then used to remove the soil and other surface materials (the *overburden*) and expose the rock beneath. This was then broken up by fires on the surface to allow metal ores to be extracted. Once the veins on the surface had been exploited in such opencast mines, the veins were followed underground. One major Roman mining area for copper was along the Rio Tinto in Spain's Sierra Morena, where mining has been underway for about 5,000 years; other important Roman mining areas included the mountains of present-day Austria and Romania. In all of these areas, and others mined during Roman times, the changes in the landscape are still apparent today. Mining has now taken place in mountains across the world, and vast amounts of ore and other materials have been removed. In some places, mountain peaks have been lowered: the summit of Cerro Rico in Bolivia, now 4,824 metres high, is thought to have been hundreds of metres higher before large-scale silver mining began there. Other mountains have been replaced by deep pits, such as the world's largest copper mines at Bingham Canyon in Utah (1.2 kilometres deep, 4.5 kilometres wide: see Figure 6); Chuquicamata, Chile (900 metres deep, 4 kilometres long, and 3 kilometres wide); and Grasberg, the world's largest gold mine (and third largest copper mine) in Indonesia. Yet other peaks have been removed and a series of terraces created: for instance, the Erzberg of Eisenerz, in the Austrian Alps, converted since 1890 into a series of terraces as iron ore has been removed; and in the Appalachians, where permits for 'mountaintop mining' for coal cover 1,600 square kilometres. One major difference between surface coal mines and ore mines is that, while the main source of waste from the former is the overburden, the proportion of desirable metals or other products from ore mines is very small, so that vast amounts of

**6. Aerial view of the Bingham Canyon Mine, Utah.**

waste rock are dumped, as are *tailings* from processing, greatly changing mountain landscapes. Such waste is, of course, also the result of underground mining. Thus, in the mountains, as across the globe as a whole, human beings are a potent agent of environmental change—even if, globally, natural processes predominate in creating and modifying mountains.

# Chapter 3
# The world's water towers

All of the world's major rivers originate in the mountains, which are thus often described as water towers: the sources of freshwater. Between a third and a half of all freshwater flows from mountain areas; billions of people rely on mountain water for agriculture, domestic use, energy, fisheries, industry, or transport; or enjoy swimming, kayaking, canoeing, or sailing on it. Both mountain and lowland people have long recognized the importance of mountains as sources of water by worshipping them as the home of deities and the source of clouds and rain that feed springs and rivers. This is true both for groups of mountains, as in the Andes where people still perform ceremonies to bring rain; and for individual mountains, such as Mount Popa, an extinct volcano in arid central Myanmar, considered the abode of powerful Buddhist spirits, or *Nats*; when people see that it is shrouded by cloud, they know that rain is likely, and they prepare their fields.

Mountains occupy only relatively small proportions of most river basins. However, the proportion of precipitation—rain and snow—falling in them is much greater because of the 'orographic effect': as air rises over the mountains, it cools, releasing the moisture it holds. This applies not only to mountain ranges, but also to individual mountains: for instance, the peak of Mount Wai'ale'ale on Kauai, the westernmost island of Hawaii, receives an average annual rainfall of at least 11.5 metres, with a record of 17.3 metres in 1982—it is the second wettest place on Earth. The greater height of mountains is important not only for triggering precipitation, but also because air temperature decreases with altitude. Consequently, there is less evaporation once the precipitation has fallen and, in many parts of the world, at least in certain seasons, the precipitation is also more likely to be snow than rain. This is probably one reason that many mountain ranges have names which mean, or include, the local word for 'white': such as the Alps, Sierra Blanca, and White Mountains. For people living in the lowlands below, sometimes thousands of kilometers away, the storage of winter precipitation as snow or ice is crucial, because when temperatures rise in the spring and summer, the snow and ice melts. The water that is released enters the rivers, flowing downstream just when it is most needed for irrigation and other uses. In total, sixty-five countries use over 75 per cent of available freshwater for food production. These include China, Egypt, and India, all of which rely heavily on mountain water, much of it from distant mountains. The river basins of these sixty-five countries cover over 40 per cent of the Earth's land surface and are home to over 50 per cent of the world's population. Equally, many of the world's largest cities—including Tokyo, Mexico City, Mumbai, New York, Jakarta, and Los Angeles—rely on upstream mountains for their water supplies.

The mountains that are most critically important for providing water for human needs are 'wet islands' in arid and semi-arid regions—often the only areas in these regions which receive enough precipitation to generate runoff and groundwater recharge. Such regions include the Middle East, South Africa, the western and eastern Himalaya, the mountains of central Asia, and parts of the Rocky Mountains and Andes. Mountains in such regions typically provide 70–95 per cent of the flow to nearby lowlands. For example, the watersheds of the Blue Nile and Atbarah rivers, which rise in the Ethiopian Highlands, occupy only 10 per cent of the Nile basin, but contribute 53 per cent of the annual inflow to Lake Nasser, as well as 90 per cent of the sediment input. The remainder of the inflow comes mainly from the White Nile, which flows from the mountains of east Africa. In central Asia, the Tien Shan and Pamir mountains occupy only 38 and 69 per cent, respectively, of the area of the basins of the Syr Darya and Amu Darya rivers which used to provide 95 per cent of the water flowing into the Aral Sea. However, during the 1960s, the Soviet Union diverted this water, mainly for irrigation to grow cotton, and this use was continued by Uzbekistan after it became independent in 1991. The losses of water through evaporation, leakage from unlined irrigation channels, and plant growth meant that, by 2009, the Aral Sea had lost 88 per cent of its surface and 92 per cent of its volume, and had split into four saline lakes. Fortunately, this trend has been partly reversed: a dam completed in 2005 has reestablished the smaller northern part of the Aral Sea in Kazakhstan.

The example of the Aral Sea shows the particular need to manage water carefully in dry regions. Yet, while mountain water is vital for livelihoods in these regions, it is also important in others. Even in temperate humid areas, mountain water contributes 30–60 per cent of the water flowing to the lowlands. For example, the Alps cover only 23 per cent of the area of the Rhine river basin, but provide half the total flow over the year, varying from 30 per cent in winter to 70 per cent in summer. It is only in the humid tropics, where the lowlands receive at least 1.5–2 metres of annual precipitation, that the contribution of water from mountain areas is insignificant. Nevertheless, it should be noted that, despite the apparent accuracy of the figures cited in this and the earlier paragraphs, our knowledge of the amounts and distribution of precipitation falling in, and the runoff deriving from, mountain areas around the world remains rather inadequate, especially in developing countries. In certain cases, this is not because the relevant data are not collected, but because governments do not want to make them available, for political reasons.

The need to have information about all aspects of the water cycle is becoming increasingly important as the world warms and human populations and their demands grow. Key issues relate not only to changes in precipitation—and whether it falls as snow or rain, and when in the year—but also to the melting of glaciers. Some supply a significant proportion of water to cities, such as La Paz and El Alto, in Bolivia, which derive 15 per cent of their water over the year as a whole, and up to 27 per cent in the dry season, from glaciers. Nearly all of the world's glaciers are shrinking, and while this process will bring more water to both mountains and lowlands for a number of years or decades, in the long term, this reliable supply of water will disappear in many parts of the world. This has already happened, for example, to the glaciers of Cotacachi in the Ecuadorian Andes, 4,039 metres high, leading to increasing local conflicts over water, which now comes entirely from annual rain and snowfall. This is a microcosm of the challenges of the future for all communities that depend to some extent on water from glaciers, whether in the Alps, the Andes, the Himalaya, central Asia, or east Africa.

## Harvesting water in the mountains

While most mountain water is used for lowland agriculture, it is also vital for mountain farmers. Irrigation systems are found in mountain areas around the world, storing water and directing it to fields at the right time and place to allow crops to grow and optimize yields. The simplest systems involve blocking streams and allowing the water to flood over meadows, which are used for grazing or hay. In more complex systems, channels bring water from high springs and streams to the fields. These systems—such as the *bisses* of Valais in the Swiss Alps, and *aflaj* in Oman (see Figure 7)—can extend for tens of kilometres, with some channels constructed around rock faces, or made of planks, suspended around cliffs. The most complex are the underground systems—*qanat* or *khettara*—a technology probably first developed about 2,500 years ago in Iran, then spread eastward to Afghanistan and

7. **Falaj Masirat Ar Rawajih, Al Jabal Al Akhdar, Oman: an irrigation channel providing water to the oasis below.**

westward through North Africa and to Cyprus. They include water collection systems, storage reservoirs and cisterns, and underground pipes that carry the water to the fields. These systems minimize evaporation, but also require high inputs of labour to build and maintain. Many, some centuries old, remain in use in Iran, the Middle East, and North Africa; others have fallen into disrepair because the required manpower is no longer available, or because pumped groundwater is more easily acquired.

A more recent technology is fog harvesting, used since the 1980s in some of the world's driest mountain areas, such as Chile's Atacama Desert. The water contained in the clouds rising over such mountains, especially in the afternoon and at night, does not always condense into rain, especially where there is little vegetation. However, erecting high fences of polypropylene mesh, or 'fog catchers', allows this water to be harvested, stored, and used for agriculture or reforestation, or piped to villages for domestic use. Installation and maintenance costs are low. Fog catchers have been installed in many other countries, including Cape Verde, Ethiopia, Mexico, Peru, South Africa, and Yemen. Again, the water provided by both of these types of technology may be endangered by climate change. Some irrigation systems are fed by glaciers and springs and, if these disappear, so will the opportunity to grow certain crops. Similarly, under a changing climate, clouds may form at higher altitudes, so that fog catchers may have to be moved uphill—which in some cases may not be possible.

## Power from water

The steep gradients of mountain rivers and streams mean that they have great potential for generating energy. The simplest technology is the water mill, developed centuries ago using local materials, and mainly used to grind grain. They are found around the world: for example, at least 250,000 grind grain in Himalayan villages. However, traditional water mills often break or are not very efficient, so that aid agencies and governments have subsidized upgrades, with wooden parts being replaced by metal or synthetic materials. A more recent innovation is to upgrade these mills to provide electricity. The simplest method is to fix a bicycle wheel to the grinding stone, so that as it rotates, the wheel drives a belt to charge a battery using an alternator. To produce more electricity—and also to mill grain—small turbines can be installed in streams and rivers. There are now over 2,000 such micro-hydroplants in Nepal. Here, and in other developing countries, the greatest challenge to making these investments economically viable is to ensure that the electricity, generated twenty-four-hours a day, is used not just in the evenings for lighting and television, but also during the day to support the local economy, for example through small-scale industries such as mechanical workshops, handicrafts, and processing of agricultural products, as well as for telecommunications. In areas with many tourists, the electricity can be used for cooking, which can reduce the demands being made on local forests for firewood. In the mountains of industrialized countries, small-scale hydropower is also expanding in response to incentives to develop renewable energy with, for example, hundreds of applications for such facilities in the Alps.

Such small-scale initiatives are at one end of the spectrum of hydroelectricity generation in the mountains. At the other end are large projects, often with sequences of large dams, such as on the Columbia River system in Canada and the USA. This is one of the world's most altered river systems, with fourteen dams on the main river and over 400 across the river basin, for both hydropower and irrigation. Globally, hydropower provides about 20 per cent of the electricity supply, in over 150 countries, and the potential for further development is huge. However, levels of development vary greatly. For example, while the Alpine countries have developed 76 per cent of their hydropower potential, Nepal and Ethiopia have developed only 1 per cent of theirs. Dams are now being constructed in these and other developing countries, often with external financing and to provide energy for export. Current rates of dam construction are particularly high in Asia, in countries such as India, Pakistan, and Vietnam—and especially China, driven by its rapidly growing economy and based on its hydro resources, which are among the richest on Earth. China is already the world's greatest producer of hydropower: developments include small decentralized plants for rural electrification as well as large projects, such as the Three Gorges scheme, providing power to urban centres and industries. In China and elsewhere, such large projects primarily benefit lowland people and economies, who gain not only electricity, but also water for irrigation, flood control, and more reliable navigation. In contrast, mountain people lose land—sometimes the most valuable for agriculture—and transport routes, and often have to be resettled: 1.2 million in the case of the Three Gorges dam. While some people close to power stations may obtain the resulting electricity at reduced prices, many do not have access to this, especially in countries in the developing world. Consequently, such large projects are facing increasing obstacles, linked particularly to social issues in developing countries, such as inadequate compensation for resettled communities; and to environmental concerns in industrialized countries, such as the fear of destroying the mountain landscapes, or losing rare and endangered mountain ecosystems or species.

## Sharing the benefits and avoiding the risks

Sharing the various benefits of mountain water has always been a challenge. At the scale of mountain valleys, irrigation systems provide good examples. These represent relatively large ventures that are constructed by the entire community, as they require major labour inputs and benefit everyone. Once constructed, maintaining the system and ensuring that the benefits are fairly distributed requires institutions and rules, which tend to increase in number with the complexity of the system. For at least 1,000 years, the elders of villages in the mountains of Oman have appointed *wakils* to maintain their irrigation channels and open the sluices according to a strict rotation to ensure the fair distribution of water. When several villages share a system, as in Pakistan's Hunza valley, each village appoints a guardian for the same purpose. In traditional societies, payment for such services is often in crops or livestock, rather than money. Similarly, water mills are also often constructed communally, and each family has the right to a specified number of hours to grind their grain. Such user-managed systems consistently outperform agency-managed systems, largely due to more effective social control and more stringent imposition of sanctions within the user communities.

Mountain communities are also developing small-scale hydropower projects, both to benefit local people directly and to provide income; in either case, community members have to agree on how to distribute the benefits, whether in electricity or in money.

While large-scale projects are developed mainly for external benefits, as mentioned in the section entitled 'Power from water', mutually beneficial arrangements can be developed between power-generating companies and communities. Two early examples of recognition of the need to compensate mountain people for providing downstream benefits were laws passed in 1916 in Switzerland and 1917 in Norway, which entitle communities to substantial annual payments and quotas of either free or cheap energy for granting the rights to hydropower development on their land (see Figure 8). In Norway, one notable beneficiary is the commune (municipality) of Modalen, north of Bergen, since three dams and four power stations have been built on its territory since 1975. As a result, the 372 inhabitants buy their electricity at half the normal price, and the commune receives about 50 million kroner (about £5 million) a year, which has been used for various purposes, including roads, free broadband for every household, a culture house, a kindergarten, a new school, a centre for senior citizens, and to support local businesses and entrepreneurship. Developing countries are also introducing similar compensation mechanisms. For instance, under Costa Rica's 1996 Forest Law, companies generating hydroelectricity from mountain water, or breweries depending on high-quality water, pay mountain landowners for the appropriate management of watersheds.

Given the dynamic processes taking place in mountain watersheds, it is not surprising that the rivers which bring benefits to both mountain and lowland people can sometimes also be sources of major disasters. One example is the semi-arid Indus Basin, which derives 80 per cent of its water from the Hindu Kush-Himalaya. This supports both small-scale irrigation in the mountains and the world's largest irrigation system, which ensures Pakistan's food supply and generates 23 per cent of its Gross Domestic Product (GDP). Every year, there are floods, particularly in the summer from meltwater and monsoon rains; over the millennia, these have brought much-needed sediment and water to support lowland agriculture. However, floods can

**8. The Oberaarsee and dam, which provides hydroelectricity to the lowland cities of Switzerland and income to local people.**

also be natural hazards, and those of July–August 2010 were particularly severe, killing 2,000 people, affecting twenty million, and causing damage estimated at $40 billion. While high levels of rainfall were certainly one factor, so were human actions, such as deforestation, drainage of wetlands, and flood control measures which had restricted the course of the river—providing a sense of security and thus resulting in the expansion of settlements and other infrastructure. Given increasing populations and demands for water, and that climate change will increase both the melting of glaciers in the upper basin and the likelihood of extreme events, many measures are needed to manage both the supply and demand of water across the basin, from the mountains to the sea.

An even more unpredictable type of event is the sudden flooding of lakes that develop on, or in front of, glaciers as they melt: such glacier lake outburst floods (GLOFs) have occurred mainly in the Himalaya, but also in other mountain ranges such as the Andes and the Alps. Although the impacts of such floods so far have been relatively limited, they could have significant downstream impacts, especially if they occur in river basins with many dams and reservoirs or large downstream populations. GLOFs have crossed the border between China and Nepal, and the Indus basin is shared between Afghanistan, China, India, and Pakistan. These are two examples of the 214 river basins, home to 40 per cent of the world's population and covering more than 50 per cent of its land area, that are shared by two or more countries. Many have assumed that such instances would lead to increasing tension and conflict between nations, especially in water-scarce basins such as those of the Euphrates, Ganges, Jordan, and Nile, which all originate in mountains. Such prophecies have not come to pass; if at all, water use conflicts occur within rather than between states. Since the early 1950s, only thirty-seven acute international disputes over water have occurred, most between Israel and its neighbours, yet over 150 treaties have been signed worldwide. However, the threat of conflicts over water both between and within states cannot be ruled out: as populations grow, demands for all uses of water increase, and climate change leads both to changes in the timing and amount of precipitation, and generally to increased evaporation. In this century, both appropriate technologies and new institutions will be needed at all scales, from villages to international regions, to ensure that the benefits of mountain water are shared, and the risks are managed, as fairly as possible, among both the guardians of the water towers and the people living downstream.

# Chapter 4
# Living in a vertical world

From a distance, or from the air, many mountains and mountain ranges look like layer cakes, with snow and glaciers icing their summits. Beneath these come the lightly coloured grasslands or tundra, then forests—typically with the darker conifers at higher altitudes, and deciduous trees below—and valleys, with rivers, fields, farms, settlements of various sizes, roads, and railways. While each of these altitudinal zones is characterized by its vegetation, they can also be described in terms of their climate, soils, and human uses, as recognized by Alexander von Humboldt two centuries ago. From his travels in the Andes, he described five major zones, each of which has since been given an English name: the lowland *tierra caliente* (below 900 metres in the Andes); the cooler *tierra templada* (montane and subalpine: 900–1,800 metres); the *tierra fria* (alpine: 1,800–3,600 metres), with a distinct cold season; the *tierra helada* (subnival: up to 4,600 metres), with snow for much of the year; and the *tierra nevada* (nival: above 4,600 metres), covered by snow for most of the year, or glaciated.

Von Humboldt recognized that climate was a dominant factor defining these zones. His model was based on his meteorological observations up to an altitude of 5,878 metres on Chimborazo in Ecuador in 1802; he was disappointed not to reach the summit—390 metres higher—of what was then thought to be the highest mountain on Earth. By 1851, he had travelled more widely, and also benefitted from the work of other explorer-scientists, and thus was able to present a map of the dominant vegetation zones of other mountains—the Alps, Lapland, Pyrenees, Tenerife, and Himalaya—in his last major work, 'Cosmos'. This showed that a second major factor is latitude: comparable zones occur at increasingly higher altitudes towards the Equator. At a regional scale, the altitude of the various zones also changes with aspect: in other words, whether mountains face north, south, east, or west. In the northern hemisphere, the zones tend to be higher on north-facing slopes because they are colder, with shorter snow-free seasons, than south-facing slopes; the reverse applies in the southern hemisphere.

This zonation is one of the main characteristics that make mountains different from other parts of our planet. The significant changes in altitude over quite small horizontal distances are comparable to a compression of latitudinal zones. For instance, in a mountain region in the tropics, one can travel from tropical rainforest to arctic-like conditions on a glacier over tens of kilometres; the equivalent of a journey of thousands of kilometres on flat land. In Europe, a rise of just 100 metres in the mountains is comparable to travelling 100 kilometres north in the lowlands. However, these are general principles and, given the great variety of interacting factors that influence mountain climates, vegetation, and soils, this simple model does not always apply. For example, not all mountains have marked zonation: on mountains near the Arctic Circle, tundra covers slopes from sea level to the summits; and very dry mountains have very little vegetation at all. There are also many tropical mountains which are entirely covered with forests; although the species change with altitude.

A further set of factors derives from past and present human uses of mountain areas. Von Humboldt recognized that each zone is best suited to particular types of agriculture. In the Andes, some have changed over the past two centuries; today, the *tierra templada* is most suited to coffee, cut flowers, and short-horn cattle; the *tierra fria* to wheat, barley, apples, pears, and dairy cows; and the *tierra helada* to rapidly developing grains, tubers, sheep, llama, alpaca, and vicuña. Here, as in many mountain regions, people—and often their grazing animals—have greatly modified the extent of each zone, creating a complex mosaic of vegetation types. In some cases, extreme human impacts have led to the loss of most of the vegetation, as on many of the mountains around the Adriatic and Mediterranean Seas from two millennia ago, as a result of forest clearing; and, more recently, in some tropical areas, after deforestation and soil erosion. One extreme example is Haiti, where mountain forests were first cleared to grow coffee in the 18th century, and have been exploited for timber since the 19th century. In 1923, the forest cover was over 62 per cent; by 1981, it was 8 per cent; and, by the beginning of this century, 2 per cent. In 2013, the government of Haiti responded to this challenge when it announced a major reforestation programme, aiming to plant fifty million trees a year, towards a target of 29 per cent forest cover in fifty years.

Dramatic changes such as significant deforestation represent the results of the incorporation of mountain areas into wider economies, leading to the breakdown of traditional linkages between mountain people and the landscapes in which they live. For most of the past two millennia, however, and even today in some places, most traditional mountain societies have tended to be based on subsistence, with limited trading for commodities that cannot be grown or found locally. One characteristic of many of these societies is that they have used the different zones of their mountains—and sometimes adjacent lowlands—in complementary ways. One example is nomadic pastoralism, still practised in many parts of Asia, including the mountains of central Asia, the Hindu Kush, Himalaya, Mongolia, and Siberia; Europe, including Norway and the Balkans and Pyrenees; and the mountains of East Africa and the Atlas (see Figure 9). Globally, sheep, goats, and

**9. Nomadic pastoralism in the Eastern Carpathians: a shepherd and his sheep in the Ciucas Mountains of Romania.**

cattle are the most important species; others include camels, horses, reindeer, and yaks. The animals are kept at lower altitudes during the winter and gradually move upwards in the spring, reaching the most nutritious high pastures in the height of the summer, when the plants have had enough time to grow after the winter snows have melted. The pastoralists and their animals then reverse the process in the autumn. However, the number of people and animals involved in nomadic pastoralism has declined over the past century, usually as the result of external forces. These include the closing of borders, such as when China closed the border of Tibet, stopping the movement of herders from India and Nepal; conflicts between pastoralists and sedentary farmers who are no longer willing to provide winter grazing in the lowlands for the pastoralists' animals; decreases in subsidies or prices—or both—for grazing animals, as in many parts of the European Union; and changes in land tenure combined with forced resettlement, often to lowland areas, as happened in many parts of the Soviet Union—although in some post-Soviet countries, this trend has reversed since 1991.

*Transhumance* is similar to nomadic pastoralism in that it involves the seasonal movement of animals from lower altitudes to high summer pastures. However, in transhumance, the owners of the animals have permanent villages at lower altitudes, where grain and hay are produced in the summer when the animals are in the high pastures; and the herders are not necessarily the owners of the animals, or even members of their families. Transhumance still takes place in many of the same regions as nomadic pastoralism, as well as in the Andes, western USA, and New Zealand, but is also declining, for similar reasons.

While many, if not most, mountain farmers own animals, the majority depend on growing crops as their main source of livelihood. However, while they are more sedentary than pastoralists, many have also traditionally used the various opportunities along altitudinal gradients by cultivating parcels of land—sometimes very small—at different altitudes to grow different crop species and varieties at each location, benefiting from the different microclimates and soils. Where soils are good, or enough manure is available for fertilization, these plots may be used for decades or longer. In contrast, where soils are less fertile, the fields are cleared, rotated over three to five years, and then abandoned for a decade or more to allow rejuvenation before being used again. This process of shifting cultivation is still used widely in many mountains in East and Southeast Asia, Central America, West and East Africa, and the eastern Andes, especially in more remote areas. The spreading of effort and risks over different altitudes ensures yields that are adequate for subsistence and, whenever possible, sale; and is typically complemented by raising animals and using diverse forest resources in various ways—again, at different altitudes throughout the year. Such patterns are found in the mountains of both developing countries, such as those of the Andes and New Guinea, and industrialized countries, such as the Alps; although increasingly practised there by part-time farmers who may no longer have the time to use their highest fields. Thus, across the world's mountains, a great variety of patterns of such vertically arranged land uses still exists, implemented by people who are not exclusively 'herders', 'farmers', or 'foresters', but who pursue a myriad activities in diverse agro-silvo-pastoral systems. However, typically, the best land for growing crops, usually around permanent settlements, is privately owned. Forests are often communally owned—unless they have been taken over by the state—and the high pastures are usually under communal management.

## Life above the trees

Even above the highest mountains on Earth, animals can survive: bar-headed geese migrate every year across the Himalaya, reaching altitudes of nearly 9,000 metres. Down on the ground, lichens establish themselves even on the highest rocky slopes of mountains, as high as 7,400 metres in Nepal. Not much further down are mosses and small plants that grow in crevices and sheltered places. Although these high places may seem very inhospitable, there is often plenty of sun, keeping ground level temperatures high for much of the day, and there is water from rain and snowmelt. The slow-growing plants that live in these areas are well-adapted to taking advantage of whatever is available during the often very short annual growing seasons, and to survive extreme temperatures, drought, and high winds. One unusual species found at high altitudes in many mountain ranges is the alga *Chlamydomonas nivalis*, which lives year-round in areas covered by snow in winter. When the snow melts, the algae germinate and multiply, creating 'watermelon snow', which is red because this green alga also has a red pigment to protect the chloroplast from the extreme visible and ultraviolet radiation at high altitudes. Many small species feed on the algae, including springtails, the animals which live at the highest altitudes on Earth: over 6,000 metres on snow and ice in the Himalaya. Many other insects, including beetles, butterflies, flies, moths, and spiders have also been found at high altitudes. Some, such as grasshoppers found at 4,200 metres in the Andes, and moth larvae found at the same altitude on Mauna Kea, Hawaii, are freeze-tolerant, but most descend to altitudes where they will not be frozen. The same applies to the various mammals that are found very high, including hares at around 6,000 metres in the Himalaya and wild dogs above 5,500 metres on Kilimanjaro.

The highest patches of vegetation in the world occur as high as 5,700–6,000 metres in the Himalaya and East Africa, with more or less continuous vegetation at 4,600–5,500 metres. This alpine tundra zone is found at lower elevations in other mountain regions, decreasing to near sea level on arctic mountains. The vegetation is dominated by a mixture of *vascular* plants—low-growing grasses, herbaceous species (herbs), and shrubs—and lichens and mosses, which tend to increase in relative proportion with both elevation and latitude. Thus, while the alpine tundra of the mountains of northern Scandinavia is dominated by mosses and lichens, there are few of these species in the alpine tundra of the Alps. Almost all of the vascular plants are perennials: completing a life-cycle in one very short growing season is almost impossible. However, annuals are found in most alpine areas, usually comprising 2 per cent of the total species or less; the proportion can rise to 12 per cent in the subtropical Himalaya, where the growing season is longer. Grasses often form tussocks, and herbs form rosettes or cushions that retain moisture, protect buds and flowering stalks, and also allow insects to survive. The small size and rounded shapes of herbs and shrubs allow them to survive in harsh conditions, and to create a microclimate that is far more suitable for growth and reproduction than would be possible if they were only a few centimetres higher. For example, the temperatures of the air above, and within, a dense carpet of azalea in the Alps can differ by 30°C.

Rounded forms also offer less resistance to the wind and do not hold the snow. In high mountains, protection from high levels of ultraviolet radiation is ensured by the angle at which leaves grow, waxy surfaces, and, sometimes, hairs on the leaves. But what we see is only a small part of the total vegetation: most of the biomass of these plants is in their roots, anchoring the plants and absorbing water and nutrients. Typically, the biomass below the soil surface is two to six times greater than above; for the sedge *Carex curvula* in the Alps, which can live for hundreds to thousands of years, the proportion can rise to eighteen times. Yet not all alpine plants are low-growing. At altitudes up to 4,200 metres in the mountains of East Africa and the Andes—where the vegetation is known as *páramo*—large plants rise 2 or more metres above the rest of the vegetation: *Espeletia* in the Andes, *Senecio* and *Lobelia* in East Africa, and the silversword in Hawaii. These strange-looking plants, with a stem surmounted by a large rosette of fat leaves that opens during the day and closes at night to protect the buds from frost, are specially adapted by having a tank root system that stores water—and they provide homes for insects that can take advantage of the microhabitats that these plants provide.

Depending on the mountain region, a wide range of animals also live in the alpine tundra. Some mammals may be permanent residents, both relatively small (such as voles, mice, ground squirrels, weasels, and marmots) and large (such as mountain sheep and goats, foxes, wolves, and bears). In winter, most of the smaller animals retreat below the ground, often to hibernate, or to live under the insulating snow. Of the larger animals, only bears hibernate; the others tend to move to lower altitudes. Many birds—large and small—also use this zone, including migrants, which benefit from the availability of insects and nutritious plants in the summer, but few stay year round, either migrating or moving downhill when temperatures drop. Today, in many parts of the world, the most frequently found animals in the alpine tundra—at least in summer—are pastoralists' grazing animals, as described in the beginning of this chapter. Traditionally, their presence was linked to small chalets or cabins in the European mountains, which used to be the only buildings high in the mountains; in other parts of the world, tents or yurts tended to be used. Since the mid-20th century, this pattern has changed significantly as the alpine zone has become a major focus of mountain tourism. In the Alps, Andes, Himalaya, Rocky Mountains, and elsewhere, resorts have been built far above the long-established settlements, for skiing in winter and a variety of activities in summer, and grazing animals have become less and less important as a source of livelihood. However, they may make an important contribution to the tourist economy: as part of the landscape tourists come to see; as the sources of local food, such as cheese; or for grazing ski runs to minimize the risk of avalanches in the winter. Even higher in the mountains, up into the nival zone, mountaineering clubs have built huts, which are used as bases for mountaineering and skiing: a phenomenon that began in the mid-19th century in the Alps, and has since spread to mountain ranges all around the world, as interest in these activities has grown and accessibility has improved.

## Mountain forests

One of the most obvious boundaries in mountain landscapes is the treeline, the upper limit of forests (see Figure 10). There have been many theories about the reasons for this upper edge, which is sometimes abrupt, but often quite ragged. Recent research in mountain ranges around the world suggests that the natural treeline for nearly all tree species is defined by the average temperature of tree roots during the part of the year without snow; if this is too low, they cannot function properly. The threshold temperature varies little around the world: 7–8°C in temperate and Mediterranean mountains; 6–7°C in subarctic and boreal mountains; and 5–6°C in equatorial mountains. Trees growing close together prevent solar radiation from reaching the ground, lowering soil temperatures. Trees growing apart from one another can survive because their roots grow into the warmer soils under surrounding tundra or grassland. Consequently, the

10. Treeline at *c*.3,100 metres on Mount Rose, Sierra Nevada, California.

natural treeline is determined by the interaction between the prevailing climate and the density of the trees. However, very many treelines around the world are not natural; they have been lowered by people cutting trees for wood and also to expand the area of summer pastures, often using fire. The process may then be continued by repeated fires or by animals—whether domesticated or wild—which graze on any new or regenerating trees, maintaining the treeline at an altitude that is lower than the natural level. The extent to which treelines have been lowered varies considerably between one mountain range and another; estimates include 200–300 metres in the Swiss Alps, more than 500 metres in the Himalaya-Hindu Kush-Karakorum, and as much as 800 metres in the Bolivian Andes—though such values are complicated by the fact that trees at the treeline are often very long-lived and may have experienced significant changes in climate over their lifetimes. In some mountains, there is almost no natural treeline, as in Scotland, where only one good example still exists, at Creag Fhiaclach in the Cairngorms, where Scots pine grows at 650 metres.

The highest forests on Earth are composed of *Polylepis* trees growing just below the permanent snow line at 5,000 metres in the Peruvian Andes. It is estimated that only 2 per cent of the original cover of these forests remains, after 10,000 years of burning, grazing by domestic animals, and climate change. In many other tropical mountains, the highest trees are of the *Erica* family, which also includes the heaths, rhododendrons, and blueberries that grow as low shrubs above the treeline in the temperate mountains of the northern hemisphere. Generally, tropical mountain forests are dominated by such broadleaved evergreen trees. These mountains may be covered in forest from top to bottom, but there are distinct forest types at different altitudes. At progressively lower elevations, both the number of species and the height of the trees increase, and there is a change from one to two to three storeys. One particular type of forest, depending on high levels of moisture, is found from 1,000–3,000 metres in tropical and subtropical mountains in Asia, North and South America, and, to a lesser extent, Africa. These tropical montane cloud forests have relatively short trees, up to 6 metres high, covered in mosses and liverworts, and are often very important both because they contain many rare species and as sources of water for people living downstream. However, they are one of the most threatened ecosystems on Earth as a result of over-harvesting, conversion to agricultural and grazing land, and other types of development.

Broadleaf evergreen trees, such as the evergreen beech (*Nothofagus*), dominate both upper and lower forests in the temperate mountains of much of the southern hemisphere. However, there are exceptions, for instance in Australia, where snow gums form the upper forests; and southern Chile and New Zealand, with both *Nothofagus* and various conifers including monkey-puzzle (*Araucaria*) and *Podocarpus*, respectively. In the temperate mountains of the northern hemisphere, conifers are also typically the trees found at the highest altitudes, with different species of pine, spruce, and fir being adapted to conditions along the spectrum from very wet to very dry. In some places, larches are present, being especially visible when their needles turn gold in the autumn. These forests also have rich populations of mosses and lichens. In other parts of the northern hemisphere, where there are few competing conifers—including Scandinavia, eastern Asia, and parts of the Himalaya—deciduous trees dominate at the treeline: mainly birches, but also alders, aspens, and beeches. However, deciduous trees are mainly found at lower altitudes, particularly in western Europe, eastern Asia, and eastern North America. Dominant genera in these multi-layered forests include ash, beech, birch, chestnut, elm, hickory, hornbeam, maple, and oak. In the autumn, when the leaves turn colour before they fall, these forests can be spectacular: 'foliage tourism' attracts up to twenty million people a year to the states of New Hampshire, Maine, and Vermont.

In total, forests cover over 9 million square kilometres of the world's mountains: just over a quarter of the global area which is covered by forests. Nearly half of the area of mountain forest consists of conifers, particularly in the middle and higher latitudes of the northern hemisphere; another 2 million square kilometres are moist tropical forests. These many types of forests are composed of a huge number of both tree and other species, which are used in many different ways. The extraction of timber can be at any scale, from the removal of individual trees for construction and other purposes, to vast clear-cutting operations, such as those of western Canada, Siberia, and Southeast Asia, to provide timber and pulp for markets that may be on the other side of the world. Mountain wood is a vital resource for mountain people around the world, not only as their most important building material, but as the primary source of energy for most of them—and also for people living in nearby urban areas, either as fuelwood or as charcoal.

Trees are not the only components of mountain forests that are important for local economies. Other products include animals, bamboos and other canes, fodder, fruits, medicines, mushrooms, nuts, and resins. In developing countries, the joint value of these products is often greater than that of wood itself. They contribute to health and livelihoods on a continual—or at least seasonal—basis both directly and through income for what can be sold, and are increasingly regarded as a key component of poverty reduction strategies. In most of the mountains of Europe, the availability of cheap wood from other sources, such as Scandinavia and the lowlands of central Europe, combined with the high costs of extraction in steep areas, means that mountain trees are barely, if at all, worth harvesting. However, other species are, including: deer and other game animals, important for hunting both for pleasure and food; and particularly mushrooms, which often have a far greater value than the trees of the same forest, and cannot grow outside the forest. For example, on average, a family in the Val di Taro of Parma, Italy earns $2,000 a year from *Boletus* mushrooms. The global trade in wild mushrooms, often from mountain forests, is estimated to be worth at least $2 billion a year; high-value mushrooms are also harvested in other mountains of Europe, western North America, and Asia. For example, in British Columbia, Canada, coniferous forests are increasingly managed for both timber and chanterelle mushrooms; in South Korea, people in mountain villages, especially those from poorer households, can earn up to a fifth of their income from matsutake mushrooms.

Mountain forests also have many other important, but less easily measured, values for both mountain communities and other people. The importance of the protection they provide against natural hazards, such as avalanches and rockfalls, has been recognized for centuries in some parts of the world; the first local regulations against cutting forests in the Swiss Alps in order to protect settlements were agreed in the 13th century. These were based on experience, after so-called 'natural hazards' had devastated settlements and crops—a pattern that, despite local regulations, recurred on ever-increasing scales in Switzerland as forests were cleared and the treeline lowered. In 1873, the first federal forestry law was passed, leading to widespread replanting and regulation of forest activities; a remarkably good investment, as it is now estimated that, without the forests, it would cost over $100 billion to ensure protection from avalanches by building permanent structures. Similarly, elsewhere in the Alps, 63 per cent of Bavaria's forests have a protective function against soil erosion and 42 per cent against avalanches. In general, the protective roles of the forests of Europe's mountains have been increasing over the past century, as these forests have increased in area not only as a result of reforestation, but also due to land abandonment. As a result of these two processes, the area of forests in Switzerland has increased by 60 per cent since the 1860s.

In contrast, the greatest rate of deforestation of any type of forest on Earth is in tropical upland forests: 1.1 per cent a year. These forests also protect against erosion and landslides, especially where levels of rainfall are high. However, although these protective roles are increasingly important as new roads are built into and across the mountains, and human populations grow, they are also often threatened as people clear forests for shifting cultivation and longer term agriculture, and as timber harvesting for both subsistence and commercial purposes expands. Consequently, the frequency and scale of natural hazards are increasing, as in the mountains of Uganda, where landslides have killed hundreds of people and destroyed important infrastructure.

Globally, the protective function has become ever more important as the number of people not only living in, but also visiting and travelling through, mountain areas has increased dramatically since the mid-20th century. One main reason for this increase in travel is the huge growth in mountain tourism; mountain forests play important roles in this as the setting for many sports, as a key element in the mountain landscape, and by protecting the roads and railways used for access. At the same time, it must be recognized that, while mountain forests do have protective roles, they may not be able to provide these when extreme events occur. For example, while the forests can protect against shallow landslides, they cannot protect against deep landslides, which are caused by tectonic movements or earthquakes that occur in many mountain areas. A recent severe example was when a major earthquake struck northeast Pakistan in October 2005, causing landslides that affected about 10 per cent of the arable land, killing over 80,000 people and affecting three to four million others. Similarly, forests may be able to withstand moderate storms but not very high winds, such as those which destroyed forests across much of Slovakia's Tatra national park in November 2004, totally changing the landscape (see Figure 11). Such extreme events are likely to become more frequent in a changing climate, underlining the need for careful management.

11. Forests in the Tatra National Park, Slovakia, where 3 million cubic metres of trees, covering 13,000 hectares, were uprooted by a windstorm on 19 November 2004.

## Valley floors and agriculture

In a mountain valley which is not settled, or visited only by hikers and hunters, there will probably be marshes, meadows, forests, and thickets of shrubs, with a wide variety of birds and other animals. Many such valleys still exist, even in the relatively densely settled mountains of Europe, Asia, and Latin America, particularly in national parks and nature reserves. However, flat valley land is the best for agriculture and settlement, so that many valley floors have been greatly changed by the clearance of vegetation, ploughing, drainage of wetlands, straightening of rivers for flood control, and the construction of buildings and transport infrastructure. As settlements grow, often taking up valuable agricultural land, their population densities may approach those of urban areas, leading to increased demands for food.

In the mountains of industrialized countries, agriculture is generally not a significant element of the economy or a major source of full-time employment: across the mountains of Europe as a whole, for instance, it only accounts for over 20 per cent of employment in Portugal and over 10 per cent in Greece, Romania, and Iceland. Many farmers are part-time, and agriculture is increasingly mechanized and supported by subsidies from national governments and the European Union. The situation is very different in the mountains of the tropics, where agriculture remains important, not least because the climates and soils in the lower mountains—from about 900 to 1,800 metres—produce higher yields than in the lowlands. Also, these elevations are healthier places to live, as malaria and other diseases are less prevalent than in the lowlands, which has led to higher population densities in the mountains. In such areas, as in Central America, New Guinea, and other parts of Asia, the good year-round opportunities for agriculture mean that there is no real need for seasonal migration, unlike in the drier and more temperate mountains.

While the good soils and availability of water mean that valley bottoms are ideal for growing many crops, land is often in short supply, especially as populations grow. Consequently—and also because valley floors do not provide the most suitable conditions for all crops—agriculture is also practised on valley sides, on permanent fields and as part of shifting agriculture where slopes are not too steep, and on terraces. In most cases, the land used for these purposes was previously forested, so that there can be risks of soil erosion and landslides if the slopes become unstable as a result of heavy rains, and especially when earthquakes occur. Such risks can be minimized by various techniques, including careful construction and management of the fields, fallow periods, the planting of trees and crops together (*agroforestry*), and the careful construction and maintenance of terraces.

Terraces are among the most characteristic and remarkable features of mountain landscapes, found around the world, sometimes on slopes as steep as 45° (see Figure 12). They take a long time to construct: 610 person-days for a 3-hectare terrace in the Peruvian Andes; 1,320 for a 1-hectare terrace on the steep slopes of Bhutan. On the steepest slopes, the walls between a terrace and the one below or above may be more than 2 metres high. Terraces are often built on slopes facing the sun, so that it can warm the soil, especially when it is irrigated; terrace and irrigation systems are often found together. Like irrigation systems, terraces are usually maintained cooperatively, although individual farmers own each one—or sometimes clearly defined parts of each—and add human and animal manure to increase fertility. The combination of warmer, wetter, and fertilized soil increases productivity, which means that terraces are often found at higher altitudes than non-irrigated fields. However, some terraces are only rain-fed, and where radiation levels are high, as in northern Pakistan, fruit and nut trees are grown to provide shade and maximize productivity. Terraces are found in the mountains of not only developing countries, but also Europe and North America, where they are particularly used to grow grapes for wine.

12. Terraces along the side of the Valle de Hermigua, La Gomera, Canary Islands.

One measure of the pressure for agricultural production is the extent to which existing terraces are used for this purpose. Where demand is very high, all terraces are in use, with irrigation occurring wherever possible to maximize yields—and there is further expansion of cultivation onto steeper and steeper slopes, with or without terracing. When demand decreases, the terraces on the steepest land and furthest away from settlements are used for grazing or gradually allowed to return to shrub and forest. Once populations really start to decline, even terraces close to villages revert to forest, as in many of the mountains of Greece and Spain, where it is quite possible that there will be large areas of uninhabited land by the middle of this century. The picture is not, of course, always so simple. Terraces may also be used less for crops, or abandoned, because the local men of working age migrate seasonally, or for longer periods, to find work. This is one of the reasons behind the decline of the intricate terraces and irrigation systems in the mountains of Yemen—many of the local men have left to work in the oil industry. Yet, even with emigration, some of the highest rural population densities in the world are found in tropical mountain areas, due both to local growth and to immigration by people from lowland areas, leading often to conflicts over land and other resources, and the need to produce ever more food. One way to do this is to expand on to ever steeper slopes, shorten rotation periods, bring in more water (if it is available) through irrigation, and apply more fertilizers (if they can be afforded). However, increased pressure on the land often leads to the depletion of soil nutrients, soil erosion, and declining crop yields.

Another way to increase production and stabilize mountain slopes is through *agroforestry*, the creation of ecosystems that are similar in structure to an area's natural ecosystems, but are composed as much as possible of species that can be used by people. One particularly good example of a traditional agroforestry system, operating at least since the 17th century, is the Chagga home gardens on the slopes of Kilimanjaro, producing bananas, beans, cardamom, coffee, onions, yams, and timber. Beehives in the home gardens produce honey that is five times more valuable than ordinary honey because of its medicinal value. Most of the farmers also have other land on the drier plains to grow staple food crops. In the Himalaya and Andes, shade trees are widely used in coffee and tea plantations. Trees that bear fruit or have leaves that can be used as fodder for animals or as organic manure can be planted along terraces, also stabilizing them. Leguminous species, such as beans, peanuts, and peas, can be planted to not only provide food, but also increase soil fertility by absorbing nitrogen from the air. Yet it is not only on the production of food that more attention needs to be focused. A considerable proportion of many crops are lost to pests, during both harvesting and storage. Consequently, better pest control, harvesting, and storage systems are also critical for increasing the food security of mountain people. Equally, many of the approaches mentioned in this section are more successful when land tenure is clearly defined; effective mechanisms exist to share and disseminate good practices and valuable innovation among farmers and through extension services; and local people—rather than government authorities or a dominant landowner—have clear political and management responsibilities and are directly involved from the outset in planning and implementing development projects.

While much of the discussion above has focused on agriculture to provide food, the growth of cash crops is increasingly important for mountain farmers in developing countries as a vital source of income. These include a very wide range of species including spices, various types of hot peppers, specialty vegetables, and flowers. All of these, as well as tea, coffee, and tobacco, are grown by individual farmers as well as in large plantations that may be owned by individuals or companies based far away; where land tenure is not well-defined or farmers can be persuaded to sell, such plantations may displace food crops. A further set of cash crops are drugs, as mountains often provide the best conditions for growing these—as well as being relatively remote areas where legal enforcement may be less restrictive. Afghanistan produces 85 per cent of the world's opium, which is also used to produce heroin and morphine. Cannabis is cultivated in mountains around the world; the Rif mountains of Morocco are the most important production area. Most of the world's coca is grown in the 'white triangle' in the Andes of Bolivia, Colombia, and Peru. All of these crops are traditionally used by mountain people, and governments rarely prosecute them for personal consumption. While mountain people gain some income from cultivating such crops, they realize few of the eventual profits, not least because they have little control over prices, and local criminals or warlords expect a share of earnings in exchange for 'security'. Drug cultivation also has environmental consequences, often including deforestation, soil erosion, declining soil fertility, and the use of fertilizers that lead to water pollution.

Other drugs grown in mountain areas are more regionally used, with less concern from the global community. The drug with perhaps the greatest overall impact on a single nation is khat (or qat), grown and widely used in the Horn of Africa and the Arabian peninsula. This requires a lot of water and, in Yemen, it has been estimated that at least a third of the national water supply goes to irrigate it. This is mainly from groundwater, and the water table serving the capital, Sana'a, has dropped significantly as a result. About 15 per cent of Yemen's population are involved in growing, selling, or transporting khat—and most Yemenis chew it. The situation is very different with betel, possibly the world's most popular stimulant, chewed by people across South and Southeast Asia. It is mainly grown by mountain farmers in the region, providing valuable incomes, and with little environmental impact as it grows on wet soils supplemented by organic fertilizers.

## Mountain settlements

Across the mountains of the world as a whole, most mountain people live in rural areas: about 70 per cent of the total population. However, there are also urban centres of all sizes in and around the edges of mountain areas. This is especially true in tropical and subtropical areas, where mountains are often preferred as places to live over the lowlands, where diseases are more widespread and the climate is less pleasant. Consequently, many of the major cities in Central and South America are in the mountains, including Mexico City (2,250 metres), one of the largest on Earth, with a population of 21 million, as well as the high capital cities of La Paz, Bolivia (3,500–3,800 metres), Quito, Ecuador (2,850 metres: see Figure 13) and Bogotá, Colombia (2,650 metres). The urban proportions of people living in the mountains of these two regions are, respectively, 46 and 55 per cent. For comparable reasons, a quarter of the mountain population in East Africa is urban. In Asia, there are also large cities within mountain areas, particularly in China, and very close to them, including two of the largest cities in the world, Tokyo and Jakarta. However, rural populations in Asia are also very large; only a fifth of the people living in the mountains of South and Southeast Asia are urban. In the industrialized countries as a whole, just over a third of the mountain population lives in urban areas. In the Alps, most of the larger cities are around the edge of the range; the exceptions are Bolzano, Innsbruck, Klagenfurt, and Trento. All of these cities

13. **Quito, Ecuador: a city of over two million people at an altitude of nearly 3,000 metres.**

are growing, with large peri-urban areas from which commuters travel daily, and are well-connected to international transport networks. In North America, there are also large cities near the mountains, particularly on the west coast and near the Rocky Mountains.

All of these large cities, as well as smaller urban centres, provide services to their rural hinterlands and are part of the global phenomenon of urbanization, drawing people predominantly from rural areas in search of employment, education, and other services and opportunities. In addition, as long as transportation is available, many people commute daily into towns and cities to work, returning to their quieter rural homes at night. In developing countries, the margins of large mountain cities are typically characterized by large densely populated squatter or informal settlements where services may be limited or non-existent; and living conditions are often better in smaller cities, which are often growing faster. During this century, more people will live in mountain towns and cities, and they will occupy an even greater proportion of the area of the mountains. Yet it is worth remembering that only a relatively small proportion of any mountain range is physically suitable for settlement—in the Alps, for instance, only a sixth—so that the majority of mountain land will always comprise fields, forests, pastures, and the summits (and sometimes glaciers) above.

# Chapter 5
# Centres of diversity

## Biodiversity hotspots

In recent decades, much attention has focused on the very high
levels of biodiversity in tropical lowland rainforests, such as those
of the Amazon and Congo river basins. Yet it is tropical mountain
forests that are the most biodiverse ecosystems on Earth. For
example, in Ecuador, 17,000 square kilometres of tropical
mountain cloud forest contain 3,411 plant species: 300 more than
in 70,000 square kilometres of adjacent lowland rainforest. The
total moss diversity of the five tropical Andean countries is
estimated to be 7.5 times higher than that of the entire Amazon
basin. The Andes are part of one of the Earth's thirty-four
'biodiversity hotspots' identified by Conservation International,
which are important not only for their high biodiversity, but also
because this is threatened. To be defined as a hotspot, a region
must have at least 1,500 vascular plants that are *endemic* (species
found in only one area) and 30 per cent or less of its original
natural vegetation. While these hotspots cover only 2.3 per cent of
the Earth's land surface, they host a very high proportion of the
world's endemic species: 50 per cent of the world's endemic plant
species and 42 per cent of endemic bird, mammal, reptile, and
amphibian species. Twenty-five of the hotspots are wholly or
partly in mountainous areas, particularly in the tropics: from
central Mexico to Argentina, through the mountains of

Mesoamerica and the Andes; the Atlantic forest of Brazil; the Horn of Africa and eastern Afromontane region, from Yemen to Malawi and Mozambique; Madagascar; India's Western Ghats; and 'Sundaland', including most of Southeast Asia, Sumatra and Borneo. Others are in subtropical and Mediterranean mountains: in California; the Chilean Andes; around the Mediterranean Sea; across Turkey and into Iran, the Cape and Karroo of South Africa; the eastern Himalaya and southwest China; and southwest Australia. Hotspots in temperate and arid mountains include the Caucasus and the mountains of Central Asia and New Zealand. In other parts of the world, such as Europe, mountains can also immediately be recognized on maps showing the areas with the highest number of species. The Alps, for example, host about 4,500 vascular plant species—more than a third of the entire European flora, of which about 15 per cent are endemic—as well as endemic insects and mammals (see Figure 14).

14. An endemic ibex (*Capra ibex*) in the Dolomites, Italy.

Many factors combine to create these high levels of biodiversity. One is the steep altitudinal gradient, combined with the fact that mountain slopes have many aspects, each with different microclimates. As a result, there is a rich variety of habitats at all scales, both horizontal and vertical: a major reason why tropical mountain forests are more biodiverse than adjacent lowland forests. This variety brings together many inter-related environmental factors: including gradients from dry upper to wet lower slopes with accumulated nutrients and debris; contrasts between sites, from very exposed to sheltered, and between shady slopes—where snow stays longer and conditions are more moist—and sunny, drier slopes; and differences in soil depth and type as well as in the degree of disturbance, for instance from avalanches and landslides. A second set of factors extends over geological time. As mountain ranges have developed, species have been able to migrate along new pathways, exploiting ecological niches as these emerge. Yet interruptions in mountain-building phases, subsequent erosion, and changes in climate—especially ice ages—have also isolated species, so that they have evolved in different ways. These two sets of factors are key reasons why mountains have such a high proportion of endemic species, which may be restricted not just to one mountain range, but sometimes to a single mountain. This applies not only to plants, but also to other species: for instance, about half the internationally recognized 'endemic bird areas' worldwide are in mountain areas, particularly tropical mountain forests.

A third set of factors leading to the high biodiversity of mountain ecosystems derives from human activities, although in different ways in different regions. In temperate regions, while the lowlands have been cleared and cultivated for centuries, if not millennia, steep and unstable mountain slopes, generally with thinner soils and cooler temperatures, provided less attractive conditions for growing crops. Consequently, these slopes were largely left alone, retaining a high proportion of their forests, although accessible areas were logged and some areas on well-watered sunny slopes were cleared to provide hay meadows. However, at higher altitudes, natural grasslands provided opportunities for grazing animals in the summer and, as described in Chapter 4, these pastures were expanded by people cutting trees and using fire. Paradoxically, the survival of many rare plant species—and associated insects—requires continued intervention in the form of mowing and grazing, so that the European Commission and the governments of its member states now pay for, or subsidize, these activities in areas where the land would otherwise be abandoned, or numbers of grazing animals would decrease below a critical level. It should also be recognized that the patterns just described are generalizations, and that there are many exceptions: for instance, the native pine forests of the Scottish Highlands had been largely cleared by 1600 AD, both for timber and to extend grazing land, and have only begun to expand again in recent decades, particularly through the efforts of conservation non-governmental organizations (NGOs).

In the tropics, people have used lowlands for intensive cultivation for centuries to millennia, but at the same time, they were often unhealthy for the farmers—although risks of disease have decreased with more efficient drainage and the use of insecticides. Yet, as noted in Chapter 4, yields are often higher in lower tropical mountains than in the neighbouring lowlands. Higher land is often also used for pastures, and fire and grazing—for 7,000 years in the Andes, and 5,000–7,000 years in the Alps and Himalaya—have had major impacts on biodiversity. As fire has been so extensive, it is usually difficult to assess how great its impacts have been. The available evidence suggests that the greater the frequency and intensity of fires, the fewer plant species are found. Trees and shrubs are replaced by dwarf shrubs and grassland, and then by tussock grasses and other species able to survive fire, such as annuals and species that have storage organs or that grow from beneath the soil surface. As in temperate mountains, appropriate levels of grazing and trampling can maintain species diversity, even when the animals eat up to 40 per cent of the plants' biomass; but excessive levels of grazing and trampling can result in decreased biodiversity, with increases in unpalatable and poisonous species, as has been found in the Pamir mountains of Tajikistan and elsewhere. Grazing animals also introduce species from lower elevations, and sometimes from other parts of the world, which are often more competitive than native species and therefore expand at their expense.

The interactions of fire and grazing are complex: grazing influences fire frequency and intensity, and fire determines how much and what species are available for grazing. Too much of either tends to lead to too many species that are unpalatable or have low nutritional value, or both—and potentially to a loss of vegetation cover, which can have other side effects, such as increased runoff and flooding. A particular set of impacts has emerged in the mountains of Australia, New Zealand, and New Guinea, where alpine grasslands evolved with grazing insects but not mammals. The introduction of sheep in Australia almost destroyed the alpine vegetation, and it is estimated that the cost of rehabilitation is twice that of all the financial benefits of pasturing, even without taking into account losses in terms of clean water and hydroelectricity. Thus, overall, maintaining high levels of biodiversity in the grazing lands of mountain areas requires a thorough understanding of the interactions of people, their grazing animals, and the ecosystems they use. Such understanding can come both from scientific research and from the indigenous, or traditional, ecological knowledge of local people, developed over generations of experimentation and experience. The diffusion and use of such understanding is essential in order to inform or establish appropriate management systems, with the active participation of all those whose livelihoods depend on grazing and the many services which these mountain ecosystems provide.

# Ensuring benefits from biodiversity

In general, as one goes up a mountain, the number of species decreases. Alpine tundra and grassland habitats, and upper mountain forests, tend to have a low diversity of genera or families, though some may include many species. Lower forests have a high diversity of both genera and families. This rich natural biodiversity is of great value to people for many purposes. Alpine ecosystems provide medicinal plants as well as grazing for domesticated animals, and for wild animals that are hunted or of value for ecotourism. As mentioned in Chapter 4, forests provide different types of wood, fibres, leaves, fruits, nuts, mushrooms, and medicinal and aromatic plants, as well as animals of all sizes which may be eaten or used in many other ways. All of these products therefore support local livelihoods, and some can provide an income if they can be sold directly to a consumer or to a trader, sometimes for export. An example of these riches is provided by Peru, a high-diversity country of which half is mountainous, where 3,140 out of 25,000 vascular plant species are used by people, including 444 for wood and construction, 292 for agroforestry, ninety-nine for fibre production, and others for cosmetics, narcotics, stimulants, dyes, and ornamentals. These numbers do not include the hundreds of species used for medicinal purposes: over 500 in Peru's northern Andes alone.

In the mountains of developing countries, there are many linkages between biodiversity, health, and livelihoods. For example, in the mountains of Nepal, where most people depend on subsistence agriculture, but also need cash for schooling and marketed commodities, there is only one doctor for every 50,000 people, but at least one healer for every hundred. Consequently, three-quarters of the population depend on about 2,400 medicinal plants for primary health care. Depending on the species, almost any part of a plant can be used for medicinal purposes. There are forty-five species of exportable medicinal and aromatic plants in the alpine zone, 114 in the subalpine zone, 225 in the temperate zone, 340 in the subtropical zone, and 310 in the tropical zone. About a hundred of these species are currently harvested for commercial use, and it is estimated that people from 323,000 households are involved in collecting them, producing up to 27,000 tonnes for export, possibly worth as much as US$30 million, which would make them Nepal's fifth most important export commodity. However, according to official records, the volume is much lower: only half, or even a quarter of this value—and only a small proportion of this goes to the mountain people who collect the plants; most is taken by middlemen. As demand has increased, particularly for export to India and other Asian countries, harvests of the most lucrative and highly demanded species have risen to unsustainable levels. For this reason, the government of Nepal outlawed the collection and trade of seventeen important species and encouraged the cultivation of thirty highly valued species. This has not worked very well, as most consumers prefer and pay higher prices for medicinal products from plants collected in the wild, and the need to earn income means that people ignore legal bans on harvesting. To ensure that local people derive a more equitable share of the value and benefits from medicinal plants will require significant effort to create the necessary structures to ensure fair access to wild plant resources—sustainable harvesting, processing, and marketing—and to implement them effectively. This should be complemented by programmes to grow species for which there is high demand and there has been over-exploitation of wild populations. However, research is needed into the best ways to propagate and grow these species, as well as to show that the medicinal value of cultivated plants is equal to, or higher than, that of wild plants. This knowledge will then have to be convincingly introduced to market chains so that consumers will buy the resulting products.

A further example from Nepal and China shows the challenges of ensuring sustainable harvesting when natural products can bring very high incomes. The fungus *Ophiocordyceps sinensis* parasitizes caterpillars, which, as a result, become mummified. The resulting combination has long been used in traditional medicine and, more recently, became widely known as an aphrodisiac. It has thus become an important source of income in rural Tibet and Nepal: in the Dolpo region of Nepal, it now provides half of the cash income and a fifth of average household incomes. Yet, harvests have been decreasing due to over-harvesting, and there have been many conflicts over access to the grasslands where the fungus lives, with several people being killed. In 2010, the Chinese ministry of agriculture established a centre for research into the cultivation of *O.sinensis*, and it is now possible to cultivate it in the laboratory. In the meantime, however, it remains an endangered species and conflicts continue.

While these examples from the Himalaya do not appear to have a particularly positive outlook, there are examples from other mountains of effective mechanisms for the sustainable use of diverse species, and equitable benefit sharing. One example is from the Kafa cloud forest of Ethiopia, rising to nearly 4,000 metres, and the original source of *Coffea arabica*—wild coffee, of which 5,000 varieties still grow there. Wild coffee is only one of many types of plants in this highly biodiverse area that are of value to people; others provide charcoal, firewood, fruit, medicine, spices, and bamboo, lianas and other building materials. Bees, which provide honey and beeswax, also live in the forests. Most of the 650,000 local people are primarily farmers, raising livestock and growing crops, including coffee that they grow in their gardens. Although the livelihoods of local peoples depend in many ways on the forests, about 43 per cent of their area was lost from 1988 to 2008 as a result of conversion to agricultural land, by logging, and for grazing, migration, and resettlement. In 2010, 760,000 hectares, including 420,000 hectares of forest, were designated a *biosphere reserve* under UNESCO's Man and the Biosphere programme, aiming to address this negative trend by finding a balance between development and conservation. The process of creating the biosphere reserve took four years, building on existing participatory forest management and family planning programmes, and involving extensive and continuous consultation of all the stakeholders: local people, government officials, and politicians. Its creation has led to many opportunities and benefits for local people. Cooperatives have been formed and currently have 6,500 members who can produce and sell more coffee, of consistently higher quality, than they could as individual farmers. The coffee is now being marketed internationally. Tourism is also being developed: hiking trails and hides to watch the abundant birds and wildlife are providing further jobs for rangers and guides. To decrease pressures on the forests for energy, fast-growing community forests have been established and 10,000 wood-saving stoves introduced. These various initiatives have not stopped deforestation, but they have slowed it—by giving people more income and job opportunities as well as by decreasing pressures on trees harvested for fuel and on rare animals and birds, which have gained added value for wildlife tourism so they are now hunted less.

## Mountain cultures

Mountain areas are centres of not only biological, but also cultural diversity, and these are often closely related. Referring to the two examples in the previous section, 102 castes and ethnic groups are recognized in Nepal; and Ethiopia's Southern Nations, Nationalities and Peoples Regional State, which includes the Kafa Biosphere Reserve, has more than eighty different ethnic groups. A further link between biological and cultural diversity is traditional ecological knowledge: for example, the use of medicinal plants in Nepal and of the many forest products used in the Kafa Biosphere Reserve, and the domestication of coffee there. In mountain areas around the world, such knowledge is vital in ensuring the survival of many plant and animal species in ecosystems which can be ordered along a gradient from 'natural' (but almost always including some intervention by people or their animals over centuries), through to shifting cultivation, agroforestry systems, pastures, and fields. Native species and varieties of plants, animals, and insects (such as bees, essential for pollinating crops) are elements of all of these types of systems, and people maintain their populations because of their diverse values.

One measure of cultural diversity is in terms of the number of languages and dialects spoken in a region, and of differences between these languages and dialects and those of adjacent lowlands. For example, there are four official languages in Switzerland—French, German, Italian, and Romansch—and many dialects of each of these, between which mutual comprehension can be very difficult. In northern Pakistan, an area described as a 'giant ethnographic museum' because of its great cultural diversity, the 35,000 people of the Hunza valley, which is about a quarter of the area of Switzerland, speak four different languages belonging to three language families. The mountainous island of Papua New Guinea is the linguistically most diverse place on Earth, with 832 living languages—so many that, for mutual comprehension, an English creole, Tok Pisin, is most widely spoken and is an official language together with English and Hiri Motu, a simplified version of the Austronesian language Motu.

Around the world, most mountain people are parts of larger cultural groups, such as the French-, German-, and Italian-speaking people of Switzerland. There are also groups who only or mainly live in the mountains, such as the speakers of the old Romance languages (such as Romansch in the Swiss Alps and Ladin in the Dolomites), the majority Burosho people of Hunza, and larger groups in other parts of the world. These include the thirty million Amhara in Ethiopia and others living in adjacent countries; twenty-six to thirty-four million Kurds, mainly living in mountainous regions of Turkey, Iran, Iraq and Syria; the nine to fourteen million Quechua-speaking people of the Andean

**15. A woman from Cusco, Peru.**

countries of Argentina, Bolivia, Ecuador, Chile, Colombia, and Peru, where Quechua is the second official language (see Figure 15); at least ten million Miao and Hmong people, mainly in China, but also in Vietnam and Thailand; ten million Uyghur in China and others in neighbouring countries; eight million Yi in China, as well as Thailand and Vietnam; 6.5 million Tibetans living in Tibet and elsewhere in the Himalaya; as well as the peoples of the Caucasus and Himalayan countries. Considering this list, it is notable that, over the past century and sometimes at present, many of these groups have been involved in armed conflict with—and sometimes experienced forced relocation, assimilation, and persecution by—the governments of the nation-states in which they live. For minority groups, maintaining identity is often a challenge to central governments based in cities far from the mountains, especially when these include significant natural resources or represent strategic frontiers—or both.

Cultural diversity can also be recognized in differences in belief systems, building materials and styles, clothing, crops, cuisines, customs, dance, handicrafts, and music. Many of the reasons for this cultural diversity may be similar to those leading to high levels of biological diversity. They include isolation, refuge from dominant cultures, use of a diversity of ecosystems, and existence far from centres of power. However, their interactions are much more complex: do people retain their specific characteristics because of their relative isolation or peripherality, or because they want to do so in order to retain their identity—or, in some cases, even to benefit by presenting a particular image to tourists? One thing is certain: no mountain group has ever been completely cut off from other people; even in the least accessible mountain areas, people have travelled out and in, usually to trade goods. In the 21st century, mountain people are increasingly being integrated into the wider world, as young people leave and return; radios, mobile telephones, videos, CDs, DVDs, TVs, and the internet become available; and tourists arrive.

The distinct clothing, festivals, handicrafts, and agricultural products of mountain people are among the greatest attractions to tourists, typically forming an important part of the 'product' marketed by governments wishing to contribute to the livelihoods of mountain people—as well as to national (or regional) budgets, and, in some cases, to maintain security (conflict is not good for tourism or the GDP). Yet in many cases, these cultural specificities lose some of their meaning, as when the timing of festivals is changed to fit with—or avoid—the main tourist season, dances that were formerly performed only once a year are performed daily during the tourist season, or festive foods are prepared for tourists on a daily basis. Traditional clothing from Thailand and the Andes has been sold to tourists in search of authentic souvenirs—and, conversely, the expectation of tourists led the villagers of Hermagor in the Austrian Alps to design a new local costume in 1965! The demand for souvenirs and local food and drink can also lead to a renaissance of traditional agricultural practices and crafts—sometimes with additional benefits in terms of maintaining biodiversity in cultural landscapes which, in turn, are attractive to tourists. It may be a challenge to ensure that the quality is high, but if it is, and the products are effectively marketed at a good price, this can provide the basis for real opportunities for employment and income. Thus, there is great potential for tourism to support and strengthen cultural identity and expression, for example in the Sherpa area of Nepal, where money earned from tourism has been invested in temples and monasteries. Mountain cultures, like their ecosystems, crops, and grazing animals, are often very distinct. However, as mountain people become increasingly integrated into the wider world, their cultures will continue to change. The great challenge is to ensure that the changes are not so abrupt that these people lose their identities and self-confidence.

# Chapter 6
# Protected areas and tourism

## Protected areas

One of the main ways that governments around the world have recognized the importance of the remarkable biological and cultural heritage of mountain areas has been through establishing protected areas, such as national parks and nature reserves. These cover almost 17 per cent of the area of the world's mountains (outside Antarctica): a much higher proportion than for the world's lowlands (11.6 per cent). At the regional scale, the level of protection varies considerably, with over 50 per cent of the mountains of northern South America protected, and high proportions also in the mountains of western North America, eastern Africa, the Himalaya, and Southeast Asia. Conversely, the lowest levels are generally in eastern North America, southern South America, and much of Africa and continental Asia. However, within every region, there are significant differences between countries in terms of the proportion of mountain area within protected areas.

Mountain areas have always been a major focus of the protected area movement. The world's first formally protected area dates from 1778, when the emperor of Manchur signed a decree to protect Bogd Khan Mountain and its sacred values. This is Mongolia's most sacred mountain, overlooking its capital, Ulaanbaatar; and in 1783, the local Mongolian government of the Chinese Qing Dynasty declared the mountain a protected site for its beauty. Nearly a century later, the spectacular Yosemite Valley was the first site set aside specifically for preservation and public use, when the US government granted it to the State of California in 1864 to protect it from settlement and commercial activity—although the homesteaders did not leave until 1874, when they received compensation to do so. In 1872, the US government created the world's first national park in the Yellowstone area of the Rocky Mountains, recognizing the area's beauty as well as the potential for tourism represented by its many hot springs and other geothermal features; the Northern Pacific Railway was already under construction and, in 1885, a spur reached Yellowstone, where travellers could stay in newly built luxury hotels. Similarly, Canada's first national park, Rocky Mountains Park, was established in 1887; the Canadian Pacific Railway (CPR) had arrived in 1883 at what is now the town of Banff, where hot springs were discovered. Ownership was contested, so the federal government took control. In 1888, the CPR opened the luxury Banff Springs Hotel.

Although indigenous people had long lived in both the Yellowstone and Banff areas, they were not involved in the creation of these national parks. They were excluded from Yellowstone from the 1870s and from Rocky Mountains Park in the 1890s, after signing a treaty with the Canadian government in 1887, permitting the extraction of mineral resources. In contrast, the origin of New Zealand's first national park, Tongariro, was the paramount chief's gift, in 1877, of these mountains to the nation for use by Maori and Europeans, to stop the land being sold to European settlers. The land was to be owned and managed in partnership; and the law to establish the national park was enacted in 1894 (see Figure 16). However, the law confiscated the summit area from the tribe's land, and it was only in 2013 that a tribunal recommended co-ownership of the national park; although Maori have been involved in developing management plans for it since the 1990s. On other continents, many of the

**16. Mount Ruapehu, in Tongariro National Park, New Zealand.**

earliest national parks were also in mountains: in Europe, Abisko, Sarek, and Stora sjöfallet in Sweden in 1909, the Swiss National Park in 1914, and Ordesa y Monte Perdido and Picos de Europa in Spain in 1918; in Africa, the King Albert National Park of the Belgian Congo (now Virunga National Park, Democratic Republic of the Congo) in 1925; and in Asia, Hailey National Park (now Jim Corbett National Park) in the Himalayan state of Uttarakhand, India in 1936.

While the concept of national parks is now well-known, having spread around the world since 1872, its origins go much further back, particularly in mountain areas, in two contexts. First, mountain people have not only recognized sacred mountains for centuries or millennia, as discussed in Chapter 1; they have also recognized special parts of the landscapes in which they live as sacred, particularly forests or groves. These are found on all continents, for example the church forests of Ethiopia, forests around the tombs and shrines of holy men in Morocco, the holy hills of Yunnan in China, and sacred groves throughout the

81

Western Ghats and Himalaya of India, in the Khumbu of Nepal, and around temples in Greece. Centuries of protection usually mean that these ecosystems are biologically more diverse and, in some cases, represent the last remnants of natural forest cover in a region or even a country. However, despite this protection, many sacred groves have lost species through over-harvesting or over-grazing. Some have become smaller, sometimes to such an extent that they are too small to provide seedlings to maintain populations of tree species; others have disappeared. A second group of places, usually of larger size, has been preserved for centuries for a different reason: to provide hunting for royalty or nobility. While many of these hunting reserves were in lowland areas, some were in mountains, for example in the Alps and Carpathians from medieval times, as well as in Afghanistan and Himalayan countries—such as Bhutan, India, Nepal, and Pakistan—from the late 19th or early 20th centuries. Strict control of poaching usually meant that populations of game animals remained high. Today, both sacred forests and hunting reserves have been formally designated by governments as nature reserves or national parks, particularly because of their high levels of biodiversity. One country where the link is particularly notable is South Korea. During the wars of the 20th century, many of the country's forests were destroyed, but those around temples survived and remained key centres of biodiversity; and so all national parks in mountain areas include temples, surrounded by sacred forests.

A more recent trend, sometimes concluding in the creation or expansion of national parks, is that of private protected areas, owned by individuals, corporations, or NGOs. Many hunting reserves, now mainly owned by NGOs, fall into this category, although sometimes the desire to stop hunting has been the impetus: one example is the Hawk Mountain Sanctuary in the Appalachians of eastern Pennsylvania, established in 1934 by the New York suffragist and socialite Rosalie Edge. About 20,000 raptors of fourteen species fly past the sanctuary each year, and it has become a major centre for research and education. In the mountains of Central America, there are many examples of private protected areas, particularly in Costa Rica, where 11 per cent of the area of national parks and 45 per cent of the area of nature reserves are owned privately. The government has issued regulations for the designation and management of these areas and also provides various financial incentives to promote conservation activities. In other countries, NGOs such as the Nature Conservancy of Canada and the Fundació Catalunya-La Pedrera in Catalonia, Spain, own large areas of mountain land. Similarly, in Scotland, tens of thousands of hectares are owned by NGOs such as the National Trust for Scotland and the John Muir Trust, as well as private individuals, such as the Danish banker Anders Holch Polvsen, the country's second largest private landowner, who manages his estates for conservation aims, rather than hunting, as is the norm in the Highlands. Globally, the largest scale initiatives have been undertaken in Argentina and Chile by Douglas Tompkins, founder of the North Face and Esprit companies. Since 1990, sometimes in collaboration with the Conservation Land Trust—an NGO which he founded—or other wealthy individuals, he has purchased thousands of square kilometres of the southern Andes. He has transferred some of these areas to other NGOs, such as the Fundacion Pumalin, which now manages both the 2,900 square kilometre Pumalin Park, recognized by the Chilean government as a nature sanctuary, for conservation and ecotourism. Tompkins has also donated land to national governments, to create the new national parks of Corcovado and Yendegaia in Chile, and to expand existing national parks: Isla Magdalena in Chile and Perito Moreno in Argentina.

Beyond the national scale, the global importance of many mountain landscapes has been recognized by their inclusion in the World Heritage List. This includes sites protected under the 1972 World Heritage Convention as having 'outstanding universal value' with regard to natural heritage in terms of aesthetics, biodiversity, or geological value (natural sites); their contributions to cultural heritage (e.g., buildings, technologies, towns, traditions: cultural sites); or as 'mixed sites'—cultural landscapes with both natural and cultural importance. Banff, Virunga, and Yellowstone National Parks, for instance, are Natural World Heritage Sites. Tongariro National Park is a mixed site because of the cultural and religious importance of its mountains, symbolizing the spiritual links between the Maori people and their environment; and Ordesa y Monte Perdido National Park is part of the Pyrénées-Mont Perdu mixed site, designated as an outstanding cultural landscape combining scenic beauty with a socio-economic structure typifying a way of life that has become rare in Europe. Designation of an area or cultural site represents the highest accolade that can be bestowed by the global community, and consequently it is notable that nearly two-thirds of mixed sites, as well as nearly half the natural sites and about one-sixth of cultural sites, are in mountains.

## People and protected areas

For about the first century of the protected area movement, the main reason for governments to designate national parks was to preserve 'natural' landscapes, often with charismatic species and attractive scenery. This approach was formalized in 1969 by the International Union for the Conservation of Nature, when it stated that one criterion for a national park was that it should contain 'one or several ecosystems not materially altered by human exploitation and occupation'. However, very few ecosystems in the world, including those in relatively remote mountain areas, have not been influenced by human activities—although the extent of modification has often not been recognized until quite recently. In other words, they are cultural landscapes. Thus, for instance, the forests of Yellowstone and Banff National Parks had been influenced by indigenous people for centuries, hunting animals and intentionally setting fires, creating a diverse mosaic of ecosystems. In the area that became the Swiss National Park, designated as an area where nature could develop without human disturbance, people had farmed and used the forests for centuries, but were required to leave their villages. King Albert (now Virunga) National Park was created primarily to protect diminishing populations of gorillas; again, the people who had used the area for centuries were moved out. The trend of depopulating places when they became national parks continued even into the 1960s, for instance when the people living in the villages of Crete's Samaria Gorge were relocated when the area became a national park in 1962. When national parks were designated in developing countries, comparable expulsions of indigenous groups often led to their near-extinction.

Since the 1980s, there has been increasing recognition that some of the characteristics for which protected areas, in mountains as in other parts of the world, have been designated require continued human intervention, rather than the exclusion of people. One example is the Tatra Mountains of Poland, where local people were not allowed to graze their sheep after the area was declared to be a national park in 1954. One result was that rare plant species on formerly grazed meadows began to decline in number because they were being shaded out by taller plants that had previously been grazed down. After this was recognized, farmers were invited to bring a controlled number of sheep back into the park in 1981, leading to the recovery of the populations of these rare plants. Similarly, in the Vanoise National Park in the French Alps, local people are now paid to mow species-rich meadows, replacing grazing animals, to maintain biodiversity. In other parts of the world, indigenous people have been allowed to continue to live, and to practice traditional resource uses, in new national parks, such as in Nepal's Sagarmatha National Park, designated in 1976.

Such trends are part of the widening, and increasing complexity, of modern concepts of conservation that recognize the imperative to address diverse needs. The first of these is to find a balance between preserving biodiversity and attractive landscapes, and fostering sustainable development. This is especially true in the mountains of developing countries, where most people still depend on natural resources for their livelihoods. The challenge is both to support livelihoods and to maintain habitats and healthy populations of species that are of importance both for local people and for biodiversity conservation—whether for their intrinsic value or, in some cases, because they are attractive to tourists and thus potentially bring income. A further challenge often emerges when wild animals, living mainly in the protected area, leave it and damage crops and kill livestock in surrounding villages, as around Nanda Devi Biosphere Reserve, in India, where leopards kill goats and sheep; and various animals—particularly monkeys and wild boar—damage crops.

This links to the second need: to involve the many concerned stakeholders in managing protected areas, rather than top-down scientifically based management by government agencies. This is particularly challenging for the majority of employees of these agencies, usually trained in the natural sciences. It represents a shift from 'expert management' to management in partnership, which requires giving away power and authority while recognizing that local people have traditional ecological knowledge and rights—both to be involved in the management of the landscape in which they live and to benefit from its resources. Such perspectives are included in recent legislation relating to national parks in some countries, such as Scotland, where the sustainable economic and social development of communities is a formal aim of the country's two national parks, which were designated in 2002 and 2003, both in mountain areas. There are also an increasing number of examples of successful co-management, often involving NGOs. These include: Fulufjället National Park in Sweden, the first in the country where local people were directly involved in designing the park and then managing it after its designation in 2002; Torngat Mountains National Park in Canada, managed cooperatively by Parks Canada and the

Labrador Inuit, who continue to use the land and resources; Jamaica's Blue and John Crow Mountains National Park, managed by an NGO, the Jamaica Conservation and Development Trust; the partnership between indigenous communities and the national forestry agency to manage the Los Flamencos National Reserve at 3,200 metres in Chile's arid *puna*; and Uganda's Mount Elgon National Park, where the national park authorities and the local parishes have signed an agreement that defines four categories of use in the park's three management zones, as well as the ways in which local people should monitor and control forest use—which they do effectively.

All of these examples include national government authorities in management structures; but there are also examples where they are not—particularly 'Indigenous and Community Conserved Areas' (ICCAs), where indigenous people or the local community are the major players in decision-making and implementation regarding the management of a protected area. ICCAs are being recognized by an increasing number of governments, whether in constitutional provisions or legislation relating to the rights of indigenous people or local communities, or in legislation relating to biodiversity conservation, often deriving from resolutions of the Conference of Parties of the Convention on Biological Diversity. Some ICCAs are based on long-established institutions, such as the Regole of the Ampezzo Valley, in the Italian Dolomites, a group of farming communities who have jointly owned much of the valley for nearly 1,000 years. In 1990, the Natural Park of the Ampezzo Dolomites was established, and the Regole has full regulatory and financial control of all of it, including the part that is state-owned. Many sacred forests and other sacred sites are also ICCAs, and the communities responsible for them have been successful in maintaining or restoring biodiversity, for example in the Western Ghats and Himalaya of India, and Yunnan and Tibet, China. In addition, in countries which recognize ICCAs, such as Mexico and the Philippines, new ones have been established from the 1990s onwards, to protect particular habitats—often with endangered species such as eagles or monkeys—and to improve the management of natural resources and ensure that local communities derive increased socio-economic benefits from them. However, many governments still do not recognize tens of thousands of ICCAs worldwide, many in mountains, and they continue to be neglected within legislation, policies, and conservation systems, and lack political and legal support.

The third need is for regional approaches, recognizing that protected areas cannot be managed as 'islands' separate from the surrounding landscape. Such approaches have developed in many different mountain regions. One important framework for such approaches is through the designation of biosphere reserves (BRs) under UNESCO's Man and the Biosphere (MAB) programme, and their subsequent implementation. Overall, a BR should be a site of excellence for exploring and demonstrating approaches to conservation and sustainable development on a regional scale. There are more than 600 BRs, and about two-thirds are in mountain areas. All BRs have one or more core areas: these must be formally designated protected areas. Around core areas are: first, one or more 'buffer zones' in which activities should be compatible with the conservation objectives of the core area(s); and, second, a 'transition area' where there should be an emphasis on sustainable resource management. A further key element of this approach is that the BR as a whole—the entire landscape, including the protected area(s)—should be managed according to a single policy or plan, steered by a structure that includes a wide range of stakeholders. This approach is being implemented successfully in many mountain areas around the world, from the Alps of Austria, Germany, and Switzerland to the middle mountains of the Czech Republic and Germany, the Atlas and Anti-Atlas of Morocco, Canada's Vancouver Island, Colombia's Sierra Nevada de Santa Marta, the Sierra Gorda and other mountain areas in Mexico, and, as mentioned in Chapter 5, the mountains of Ethiopia.

As many mountain ranges are also national frontiers—and animals, birds, people, and pollution cross these—some regional approaches have to be transnational. These include six of the world's transboundary BRs, shared between the Czech Republic and Poland; El Salvador, Guatemala, and Honduras, France and Germany; Poland, Slovakia, and Ukraine; and Portugal and Spain. These are all examples of *connectivity conservation*: regional-scale approaches in which protected areas and wider landscapes, and the people who live in them, are linked in order to foster both biodiversity conservation and sustainable development. Such transnational initiatives have been developed in many parts of the world, including 'Yellowstone to Yukon' along the Rocky Mountains of Canada and the USA; the Mesoamerican biological corridor from Mexico to Panama; 'transboundary landscapes' across the Himalaya and Hindu Kush, involving China, India, Myanmar, Nepal and Pakistan; from Spain's Cantabrian mountains through the Pyrenees to the Alps; between Lesotho and South Africa; and in the Altai, shared by China, Kazakhstan, Mongolia, and Russia. Interestingly, many of these have often not been initiated by governments or their agencies, although they have generally been key players.

## Centres of tourism

While the protection and management of areas with high biodiversity value and, often, attractive scenery are typically the formal reasons for which governments establish protected areas and propose sites for inclusion on the World Heritage List, a further, often unstated, reason is usually the potential for bringing in tourists, especially to regions that are relatively remote and where economic development has been identified as an imperative. This has been evident since the establishment of the first national parks in North America over a century ago, and remains a key challenge, particularly in developing countries such as China, which has created many national parks in mountain regions since 1982. These are 'self-funding organizations', which have to generate revenue to support a range of activities including economic development, infrastructure construction, poverty relief, and tourism (see Figure 17). However, there is generally little emphasis on long-term planning and monitoring, based on scientific criteria. The need to generate income, especially through

17. The Bailong elevator, in Wulingyuan World Heritage Site, China, is 330 metres high, and was opened to the public in 2002.

tourism, has led to the deterioration or even loss of many of the habitats and species for which the parks were designated, as well as serious water pollution problems and, in some cases, conflicts with local people.

China is one of the many emerging and developing countries where the growth of tourism is particularly rapid. Tourism is one of the world's largest and fastest-growing industries, and it has been estimated that, of its global receipts—over US$1,000 billion a year—up to 20 per cent is associated with travel to mountain areas. The Alps are one of the world's major tourist destinations, attracting about ninety-five million long-stay and sixty million day visitors each year. The latter figure may be related to the fact that, around the world, most tourists visit destinations in their own region; and over half of the world's international arrivals are from Europe.

While global mass tourism is a phenomenon that emerged after World War II, the roots of the industry are in pilgrimage, a centuries-old phenomenon that is still important in many mountain areas. Every year, millions of pilgrims still continue to visit *Dev Bhumi*, the land of the gods, in the Himalaya of the Indian state of Uttarakhand, which contains many temples and shrines. Tourism accounts for a quarter of the GDP of Uttarakhand, and about 70 per cent of the tourists are pilgrims. Over 900,000 come each year to Badrinath, at an altitude of 3,300 metres, where the Hindu temple has been a site of pilgrimage since the 9th century AD. The number of pilgrims has grown ten-fold since the Indian government built a road into this area after the 1962 war with China. As a result, most of the people who had provided food, accommodation, and transport along the formerly arduous pilgrim trail lost these additional sources of income. In Badrinath, the massive influx of tourists led to serious sanitation and waste management problems, and the destruction of a sacred forest. After failed attempts by the Indian army and forestry department to plant trees, the chief priest developed a partnership with local scientists, who provided seedlings. Pilgrims have now reestablished the sacred forest and local people continue to maintain it—and gain blessings for doing so. While pilgrimage is still an important component of tourism in Uttarakhand, sacred sites in other parts of the world are now visited by vast numbers of people whose motives for visiting may be only slightly religious at most. These include Jebel Musa or Mount Sinai in Egypt, a destination for day trips from the Red Sea coast; T'ai Shan in China, a mixed World Heritage Site with a cable car to its summit, although most visitors climb the 6,600 steps; and Mount Fuji in Japan, climbed by 300,000 people a year and designated a Cultural World Heritage Site in 2013—with a requirement for a detailed management strategy to stabilize the trail network, manage delivery of supplies and energy, and increase the environmental awareness of visitors.

The example of Badrinath shows many of the challenges associated with the development of tourism in mountain areas, but also that innovative solutions can be found. For any destination, access is a critical factor in the development of tourism; roads built for other reasons, such as the extraction of natural resources or military purposes, often lead to the introduction of tourism to remote mountain areas and its subsequent increase. Historically, the development of tourism in many mountain areas was associated with the advent of railways. The first railways were built into the Lake District in the 1840s, and the Swiss Alps in the 1850s. This trend was not limited to Europe: in India, railways to 'hill stations' were built from the 1850s and until the end of the century; transcontinental railways were completed across North America from 1869 to 1885, crossing the Rocky Mountains and other ranges; and the Trans-Siberian railway was completed in 1909, providing access to many remote mountain areas, although few have yet been developed for tourism to any great extent. Since the 1960s, high-speed trains in Japan, Europe, China, and South Korea have linked major cities to tourist destinations in the mountains. In addition, in recent decades, a key driving force of tourism to mountain areas around the world has been the decreasing real costs of international air travel; and, within countries, regional airline companies and, in some cases, helicopters make it possible to reach almost any mountain area.

In addition to access, the other crucial driving forces leading to the growth of tourism in mountain, and other, areas in recent decades have been increases in the time, discretionary income, and mobility of the expanding population of an increasingly urbanized world—and especially the desire of a growing middle class to escape from cities for recreational and spiritual needs. Many governments and communities in mountain areas around the world have come to regard tourism as vital for economic development, and even survival. Yet its distribution is very uneven in every mountain region, and its benefits tend to be spread very unevenly at every scale, from the national to the local. At the scale of communities, the apparent benefits of tourism in terms of maintaining populations are shown by statistics from the Alps, which generate nearly 50 billion euros in annual turnover, about 8 per cent of the annual global tourism turnover. Yet, while tourism provides 10–12 per cent of the jobs in the Alps, tourism-related activities are concentrated in only 10 per cent of the communities. Generally, these have stable or growing populations, while other rural communities are losing population. However, even in this global centre of tourism, demand is not reliable; in the 1990s, the number of overnight stays decreased significantly, although demand has since recovered, particularly in the Austrian and Italian Alps. Tourists are also highly sensitive to real and perceived risks to personal safety—as seen in the mountains of Afghanistan, Colombia, Kashmir, Nepal, Pakistan, Rwanda, and Yemen in recent years—although tourists may begin to return quickly when the perceived risks decrease, bringing much-needed income. A further set of challenges can emerge after major natural disasters; for example, the June 2013 floods in Uttarakhand destroyed not only villages, but also many tourist facilities, roads, and 145 bridges, causing billions of dollars' worth of loss to the tourism sector, which will take many years to recover.

Mountain tourism comprises a massive and complex set of interactions of people involved in a vast array of subsectors. *Ecotourism*, with the potential both to conserve key habitats and species and to support local economies, is growing particularly fast, often connected to *agritourism*, which gives added value to cultural traditions and can be associated with the renaissance of traditional products, especially those linked to farming. Wellness tourism has a long history in the many mountain areas with geothermal resources. There is a vast range of types of sport tourism, which are continually evolving and being introduced to new locations; each is typically linked to the development and marketing of new technologies—such as carving skis, lightweight snowshoes and paragliders in recent years—and the related, often 'extreme', experiences. Each subsector involves a different clientele in a highly competitive, unpredictable, and increasingly global market. Each type of visitor is attracted by different characteristics of mountain environments, landscapes, cultures, and/or possible experiences, and demands a specific range of services and facilities. They also have different expectations and ways of using mountain landscapes, which can lead to conflicts, for example between hikers and mountain bikers.

Changes in fashion, as well as in the expectations of tourists, often mean that investments made to attract one type of tourist have to be supplemented by investments in new facilities, to either encourage repeat visits or attract new types of tourist. A further challenge is that, despite the great diversity of mountain tourism opportunities, most activities are also seasonal, especially if they are conducted out of doors; this is also influenced by the holiday periods of the main sources of tourists and variations in the ease of access throughout the year. Except on artificial slopes, skiing is only possible in cold temperatures, relying on natural and, increasingly, on expensive artificial snow. Many types of ecotourism are related to the annual cycles of particular plants and animals. People who want to camp, hike, or climb prefer the seasons with the least rain and fewest annoying insects. This seasonality presents many challenges in terms of business continuity and employing high-quality staff. Although most resorts are trying to lengthen their season, for instance through offering low rates for conferences or organizing festivals in the off-season, services and facilities required by tourists in one season may not be appreciated by those who come at other times. For instance, people who come to enjoy the summer landscape do not enjoy seeing ski lifts and other evidence of winter activities. Marketing is essential to attract tourists; but the image they leave with, and communicate to their friends, may be just as important in ensuring future visits to any tourist destination.

## Positive and negative impacts

Tourism is part of the process of the globalization of mountain areas, even though the majority of visitors to most countries are domestic rather than international; the few exceptions are countries such as Andorra, Austria, and Nepal, which are small and/or adjacent to countries with much larger populations. Tourism can provide important new sources of income for local people who provide goods—such as food, handicrafts, or sports equipment—and services, such as accommodation, guiding, portering, or restaurants. At the individual level, a few people get rich from tourism: often immigrants or others with access to outside capital. However, the financial success of entrepreneurial individuals or companies may depend on the majority of the workforce who work for relatively low wages and may not even be able to find affordable housing in the communities where they work, unless their employer provides it. Both of these problems are compounded by the seasonality of tourism in most mountain locations, which means that it cannot be a year-round source of employment for many. As tourism becomes dominant in the local economy, the costs of food, goods, services, and places to live tend to rise. Nevertheless, all of these challenges can be addressed by governments, companies, and individual employers. Governments, in particular, can facilitate good practice through financial measures, such as subsidies or loans for businesses based in mountain areas, incentives to encourage local products, or regulations relating to employment, planning, or even the cost of staple commodities. There are many guidelines for sustainable tourism produced by international organizations—such as the World Tourism Organisation and the United Nations Environment Programme—and trade organizations which are being implemented by companies. Participatory and well-planned tourism development can lead to benefits for a large proportion of the local population, including women, whose status may increase as a result. Tourists also expect healthy living conditions; this can lead to public health benefits for local people if local authorities invest the profits from tourism wisely.

While the benefits and disadvantages of tourism for local economies and employment are often hard to predict and assess, it is even more difficult to anticipate the inevitable cultural changes. Although the costumes and traditions of indigenous people are often among the reasons that tourists visit, these often change as 'Western' clothing and footwear are adopted—often as status symbols, especially among young people—and as cultural activities are adapted to the demands of tourists. To avoid such impacts, some communities, such as the Zuni of New Mexico, have decided that the disadvantages of tourists outweigh any potential benefits and have excluded them from their festivals. Similarly, while the desire of tourists to take souvenirs home can lead to the revitalization of local skills, souvenirs may be imported from far away for sale to tourists, and priceless cultural artifacts may be stolen and sold. The physical changes resulting from tourism can also extend to design of buildings and settlements. In mountain regions with a long human history, the characteristic design and ornamentation of traditional buildings in each valley, town, or village makes them potential attractions for tourists. However, the introduction of tourism typically leads to homogenization and to the construction of new buildings—and even settlements, including many ski resorts—that are often not particularly well adapted to the rigours of the mountain climate in terms of design, construction, and energy efficiency. Nevertheless, there are also examples of places where design has been sympathetic to the site or new technologies have been introduced while retaining traditional design—such as the use of passive solar technology in Ladakh, in northern India—so that resources are used more efficiently to the benefit of both tourists and residents. While many mountain tourists, especially those most interested in sports activities, may not expect or particularly care about authentic food, buildings, or cultural activities, an increasing number do—particularly within the growing ecotourism and 'responsible tourism' market, in which the value of diversity is explicitly recognized. Thus, while tourism can have many negative cultural impacts, it can also lead to the rejuvenation of traditional skills, providing employment, especially in the off-season, and also assisting in strengthening of cultural identity.

Tourism can also bring negative environmental impacts, both locally and over a broader region. Local impacts mainly occur along the valley floors and on lower slopes, where most infrastructure is constructed: they may include the loss of agricultural and residential land; air pollution, particularly from traffic and in inversion conditions; and water pollution from inadequately treated waste and badly constructed roads. Road construction can also lead to increased runoff and erosion. As tourists visit higher and more dispersed locations, erosion occurs along trails and around campsites, requiring restoration; and levels of waste and faecal water pollution often increase. Extreme examples are the 23 tons of waste found along the 53-kilometre-long trail around Mount Kailash in 2013, and 12 tonnes a year of excrement collected at Mount Everest Base Camp. These are particular challenges as waste decomposes slowly at such a high altitude—although a solution may have been found for the latter: the world's highest biogas digester, which would produce energy for local people.

In the mountains of developing countries, both tourists and those providing services to them often use trees and shrubs for cooking and heating, leading to changes in forest composition and structure, and sometimes to their loss. In downhill ski resorts, there are impacts up to the summits of mountains, through the logging and bulldozing of ski runs; the construction of chairlifts and cableways; the installation and use of snow cannons; and the disturbance of vegetation and wildlife by skiers and machines. Yet, while all of the impacts mentioned in this chapter have been recorded in mountain regions around the world, there are many means for addressing them through effective and often innovative technologies and strategies—for example, in energy production and utilization, water and waste treatment, traffic management, and trail and road design and construction—as well as guidelines on how to implement these, and regulations to ensure this happens. There is also a vast range of accreditation and certification schemes and awards that aim to foster tourism that brings environmental, community, or cultural benefits. Some destinations—such as the resorts of Åre in Sweden, Aspen-Snowmass in Colorado, Park City in Utah, and the twenty-seven members of the Alpine Pearls Association across the Alps—are actively focusing on such issues, and marketing their sustainable credentials to potential visitors.

To prosper in this competitive industry, those involved in tourism in every mountain community and region need to develop a unique image based on local environmental and cultural assets—one in which the income from tourism should be invested to ensure a sustainable future, even if numbers of tourists decline. Most types of mountain tourism are seasonal and unpredictable in the long term, and therefore it is essential that its development is linked to that of other economic sectors. The rapid development of tourism typically creates great demands and stresses on local families, communities, and infrastructure. Rather than shifting too far into providing goods and services for tourists, people working in tourism need to maintain other opportunities for employment and income, whether in agriculture, forestry, industry, handicrafts, tele-working, or commuting in the off-season. Local, regional, and national governments need to develop and implement policies and financial instruments to ensure that the economic benefits of tourism—both locally and to national economies—are reinvested in the resources that attract tourists. These include not only physical infrastructure, but also landscapes and cultures. Equally, governments and development agencies need to provide the resources to train mountain people in skills necessary both in tourism and for other employment opportunities. Mountain tourism will remain a vital part of the economies of many mountain communities, and important for some national economies but, for most, its long-term future will remain uncertain, especially as the world's climate changes.

# Chapter 7
# Climate change in the mountains

## Evidence from the mountains

Life on Earth is only possible because of the existence of 'greenhouse gases' in the atmosphere which allow sunlight to reach the Earth's surface but absorb the infrared energy that is emitted. While the most abundant and important greenhouse gas is water vapour, human activities have only a small direct influence on its overall concentration, and its *residence time* (the time it stays in the atmosphere) is only nine days. The two other major greenhouse gases are carbon dioxide and methane, and their concentrations in the Earth's atmosphere have been increasing since the beginning of the Industrial Revolution in the mid-18th century, due particularly to the burning of fossil fuels, the clearing of forests, and other changes in land use. Molecules of carbon dioxide and methane stay in the atmosphere for years after they are emitted: their residence times are, respectively, five to 200 years and twelve years—so human activities have long-term effects.

The clearest direct evidence that concentrations of carbon dioxide are increasing around the world comes from continuous measurements taken since March 1958 near the summit of Mauna Loa, Hawaii, at an altitude of 3,397 metres. The site was chosen by the US Weather Bureau after it was unable to find anywhere in the continental USA where the air was clean enough. Charles

Keeling's original reason for monitoring carbon dioxide was to better understand the diurnal cycle of its concentration in the atmosphere. However, within a few years, he had identified a seasonal cycle—because vegetation in the northern hemisphere absorbs carbon dioxide in summer and releases it in winter—and also discovered that, each year, concentrations were higher. Since 1958, the concentration of carbon dioxide at Mauna Loa has increased from 317 to nearly 400 parts per million. Thus, the tallest mountain on Earth, far from any industrial activity, is the location where the greatest challenge to the future of mankind became evident: there is now a global consensus that the Earth's climate is changing because concentrations of carbon dioxide and other greenhouse gases are increasing as a result of human activities.

Mountains are not only where the primary cause of climate change has been measured, but where crucial evidence of its effects can be seen. Two of the most compelling types of evidence are the melting of glaciers and the upwards movement of plants on mountain summits. An international programme of glacier monitoring began in 1894, as it was hoped that long-term observations would provide insights into processes of climate change, such as the occurrence of ice ages. Initially, the programme focused on measuring the lengths of glaciers. Subsequently, the monitoring programme was expanded to include the *mass balance* of glaciers—the difference between the inputs of snowfall and losses, mainly by melting—and the use of images from satellites. The great benefit of these images is that they provide complete coverage of areas which are often not easily accessible, such as many glaciated areas. Since 1986, the resulting standardized data on changes in glaciers, provided by scientists in over thirty countries, has been compiled by the World Glacier Monitoring Service. These data show that the world's glaciers are shrinking—and at an accelerating rate. While the primary reason is a warming atmosphere, an additional cause, also of human origin, is the deposition of fine particles of black carbon, or soot, on the surface of glaciers. This comes mainly from engines and the burning of coal and other biofuels. Because soot absorbs more radiation than ice, its presence increases rates of glacier melt. This phenomenon occurred in the Alps in the 19th century, so that this was the region where initial rates of glacier melting were greatest, and is currently happening in the Himalaya.

Over the past century, more than 600 glaciers worldwide have disappeared. A few glaciers have advanced in recent decades—for example, where snowfall increased along the west coasts of Norway and New Zealand until the late 1990s, and also in the western Himalaya—but these are exceptions. At the global scale, the greatest losses of glacier ice from 2003 to 2009 were in Alaska, the Canadian Arctic, Greenland, the high mountains of Asia, and the southern Andes. Together, these accounted for 80 per cent of the total loss of 260 gigatonnes per year. In terms of area lost in particular regions, the glaciers in subtropical and tropical mountains have been particularly affected. For instance, over the past three decades, the total glacier area of Nepal decreased by 24 per cent, at an average rate of 38 square kilometres a year; the estimated ice reserves shrank by 29 per cent—129 cubic kilometres. In the Andes, where most of the world's tropical glaciers are located, the rate of loss of Peru's glaciers more than trebled from 1964–75 to 1976–2010, and those of the Cordillera Blanca lost about a third of their area—200 square kilometres—from 1980 to 2006. In temperate mountains, the glaciers of the Alps are losing 2 to 3 per cent of their area and volume each year (see Figure 18); and in the USA, the number of glaciers in Glacier National Park, Montana, decreased from 150 in 1910, when the park was established, to only twenty-five larger than 10 hectares in 2010. Further north, the 300 glaciers of Iceland, which is warming at four times the average rate for the northern hemisphere, are losing about 11 billion tonnes of ice a year.

While long-term measurements of carbon dioxide and glaciers were planned, finding evidence of the upward movement of alpine plants was largely serendipitous. In the first two decades of the 20th century, the eminent Swiss botanist Josias Braun-Blanquet,

18. The author in the Morteratsch valley, Switzerland, in October 2013. Fifty years earlier, he had stood in an ice cave in the snout of the glacier at this location. The glacier is now more than a kilometre shorter; it can be seen in the distance.

a keen alpinist, climbed many peaks in the Swiss Alps. While on their summits, he recorded the highest specimens of each plant species. In 1957, he proposed that these high summits could be good locations to monitor the effects of climate change because, like Mauna Loa, they are relatively distant from direct human influence, so that any change can be assumed to be natural in origin. In the early 1990s, two young Austrian botanists, Michael Gottfried and Harald Pauli, returned to the peaks that Braun-Blanquet had climbed, and repeated his measurements. They found a statistically significant increase in the number of species on the summits. These changes could then be correlated to changes in regional climates. In 2000, this research led to the creation of a global research network, the Global Observation Research Initiative in Alpine Environments (GLORIA). The occurrence and altitude of plant species are now being measured using standardized techniques on the summits of mountains in 121 locations in forty countries on five continents. The first results, from both Australia's Snowy Mountains and sixty-six European mountains, showed significant changes in plant species diversity over a decade. However, while numbers of species increased on summits in Australia and in northern and central Europe, the numbers on Mediterranean mountains were either constant or decreased, probably because of drier and warmer summers. This suggests a serious threat for the future survival of these species, as such climatic trends are likely to continue.

## Changes in mountain climates

Since the late 19th century, the Earth's average temperature has increased by about 1°C. This trend has not been continuous, and there have been considerable variations between regions. It is also important to note that the data on which such statements are based are obtained from locations that are not uniformly spread around the globe. There are far more, and longer, records from industrialized countries. In mountain areas, most measuring stations are on valley floors, and relatively few are above 500 metres, with very few above 2,000 metres. Despite these limitations, the available data from the few mountain regions with large numbers of measuring stations that have long records show that these regions have warmed more rapidly than the global average over the past century. The Alps warmed at double the global rate, and temperatures at higher elevation sites in the Rocky Mountains in Montana, Wyoming, and Idaho rose at three times the global average. For other mountain regions, records from individual stations on all continents except Antarctica generally show increases, although they cannot be assumed to represent the regions as a whole. With regard to precipitation (rain and snow), past patterns of change have been far more variable than for temperature. These patterns are particularly complex in mountains, as precipitation is strongly influenced by local topography and, in many areas, a significant proportion of the annual precipitation falls as snow rather than rain—and measuring snow is usually less accurate. Nevertheless, the available long-term data show that, over the past two centuries, precipitation generally rose in the northwestern Alps and decreased in the southeastern Alps. In recent decades, as winters have warmed, there has been less snow and more rain: a pattern also found in other mountain ranges, including those of the western USA and Canada, the western Himalaya, and Japan.

Looking ahead, different global climate models provide contrasting results for future temperatures and, especially, precipitation; and the rates of change will depend at least partly on policy decisions and other human responses to climate change. Predicting the future climates of mountain areas is especially challenging, given our limited understanding of their recent and current climates. However, it is likely that many mountain areas will continue to warm more strongly than the global average, especially at the highest altitudes, and particularly in winter. Although total precipitation may increase in some mountain areas, such as those of northern and central Europe, the Rocky Mountains, and western New Zealand, there are likely to be other areas where there is a decrease, including the Mediterranean and many arid regions. The proportion falling as snow is likely to decrease in most places.

Although average values of temperature and precipitation are relevant, extreme values are often more important. People can adapt to gradual changes, but are often unable to cope with extreme events, such as very heavy rainfalls, leading to floods and landslides; very dry periods, leading to fires and the loss of crops and animals; high winds, toppling trees and damaging infrastructure; or heavy snowfalls, causing avalanches. All of these extreme events are likely to increase in frequency and magnitude in mountain areas. Particularly critical may be an increased frequency of hurricanes, affecting the mountains on Caribbean islands and Central America, most of which are highly vulnerable because their volcanic soils and steep terrain mean that they are prone to mudslides—the main cause of casualties and destruction. In Asia, more severe monsoon rainfall is likely, potentially affecting tens or hundreds of millions of people.

## Impacts on ecosystems

All of these various changes will affect not only people, but the ecosystems on which they depend both directly and indirectly. In the Andes, the loss of glaciers will affect the wetlands that depend on their meltwater, providing vital habitat for both permanent and migrating species. As climates warm, mountain species may be able to move to more appropriate new habitats more easily than species living at lower altitudes, because the steep topography means that the distances they have to move are generally smaller. As the GLORIA programme is beginning to show, alpine plants can move upwards as the climate warms. Similarly, trees are moving to higher altitudes—for example, oaks and beeches in Spain's Montseny mountains—and treelines in mountains around the world are advancing upslope. More mobile species have already begun to do so, including reptiles and amphibians in Madagascar; butterflies in France, Spain, and the western USA; birds in Costa Rica; and pygmy rabbits in the western USA. However, a major challenge for such upward movements is that the area of suitable habitat decreases at higher and higher elevations: some species may eventually have nowhere to go. This is a particular challenge for those living at the highest altitudes on tropical mountains, where not only changes in climate, but also decreases in the availability of moisture as clouds form at higher altitudes, and increases in the range and virulence of pathogens, may affect chances of survival.

Some species have already become extinct, most notably amphibians living in the cloud forests of Costa Rica. In 1989, the last golden toad (*Bufo periglenes*) from the Monteverde Cloud

Forest Reserve was seen; and of the many harlequin frog species living in the mountains of Central and South America, two-thirds have disappeared in recent decades. Changes in climate may have been the direct cause; another possibility is that warmer nights and increased cloud cover may have created optimum conditions for the growth of a particular fungus that kills these species. There have also been regional extinctions, for instance of many colonies of pikas, small mammals that live in the alpine zone of mountain ranges in the western USA, because warmer climates have led to the loss of sufficiently cool places for them to live. Drought and high summer temperatures have led to the extensive death of eucalypts in the subalpine zone of Tasmania's mountains, and trembling aspen and lodgepole and pinyon pine in western and southwestern North America. To date, extinctions of species have been rare, but changes in climate can also affect the survival of populations. For example, female Columbian ground squirrels in the Canadian Rocky Mountains now emerge from hibernation ten days later than they did twenty years ago (see Figure 19). The key factor is late snowmelt: more frequent late-season snowstorms have meant that snow remains on the ground longer. As a result, the animals have less time to accumulate the fat they need for hibernation, and the population is not growing as fast as it used to, affecting its long-term viability. Conversely, the impacts of a changing climate may also be positive: for example, at 2,900

**19. Columbian Ground Squirrel in Glacier National Park, Montana, USA: one of many high-mountain species affected by climate change.**

metres in the Colorado Rockies, snowmelt is now two weeks earlier than in 1975, and marmots are able to put on more fat before entering hibernation.

While a suitable habitat is essential for species to thrive, other factors may also be vital. Plants that depend on pollination for reproduction require that the necessary pollinators are available; predators require prey; newly hatched birds need food; and herbivores need the correct plants to eat. However, the different species in these inter-specific relationships may respond in different ways to changing climates—especially if one species lives year-round in the mountains, and another migrates there from other locations where the climate may also be changing. For example, in the Colorado Rockies, the nectar-producing plants used by broad-tailed hummingbirds migrating from Central America are flowering earlier, so that by the time the birds arrive, there is less nectar available for them. If this trend continues, the birds may not have enough time to nest and fledge their young. In Scotland, golden plovers migrating from the lowlands to the Highlands require cranefly to feed their chicks, so if this food is not available, they may not grow and survive. This is a complex phenomenon, as the abundance of cranefly depends on how the previous summer's weather has affected their larvae. In the Qinling mountains of China, populations of giant pandas may be threatened by reductions in the range of the three dominant bamboo species, their main food source. And as forests replace open habitats in the Hindu Kush-Himalaya, 30 per cent of the snow leopard's habitat may be lost.

Over this century, the complex interactions of climate and species mean that more species are likely to go extinct, but also that new ecosystems may emerge as new species move into the mountains from lower altitudes and interactions between species change. For plants, changes in snowcover duration and the length of growing season are likely to have greater effects than direct effects of temperature on metabolism. Species currently living at higher altitudes may be particularly threatened, both because there is less habitat above them and because they will not be able to compete with species arriving from lower altitudes. Of the four trout species living in streams in the western USA, all will lose habitats, but the degree to which populations are affected will depend on interactions between higher temperatures, flow rates, and populations of other species. However, for most species in most mountains, the various interactions are often not well understood, particularly because of a lack of long-term monitoring; the examples from the Colorado Rockies mentioned in this chapter are atypical, based on fieldwork at the Rocky Mountain Biological Laboratory since 1975. To plan ahead in order to maintain populations of many species requires such long-term data. In addition, regional approaches to biodiversity conservation, such as connectivity conservation, will become more important, as climate change is only one of many factors influencing the distribution of species: changes in land use will also continue to influence very many species and habitats.

## Challenges and opportunities

Losses in the number, area, and volume of glaciers are clear expressions of the impact of changing mountain climates—and these trends will continue. For years to decades, the increased runoff may benefit the cities, industries, and famers utilizing glacier meltwater directly. Globally, 140 million people live in river basins where at least 25 per cent of the annual flows come from glacier melt; 370 million for a threshold of 10 per cent. Globally, 90 per cent of the population at risk lives in Asia. However, once glaciers have gone, only seasonal rainfall will be available. This has already happened for some small settlements in the Andes and is likely to become a reality for the capital cities of La Paz (and neighbouring El Alto), Lima, and Santiago de Chile, all of which depend to some extent on glacier meltwater. For most of the world's population, whether they live in or outside the mountains, changes in the timing and amount of precipitation will be more important, as will be the proportion of precipitation falling as rain rather than snow. This will affect water supplies not only in, but also far from the mountains, because rain flows downstream immediately, while the water in snow is stored until the spring melt. Consequently, the maximum runoff will be earlier in spring, or even in winter, and there may be increased numbers of floods, particularly when rain falls on frozen ground. Conversely, where less water is available in summer, as is expected over much of Europe, this may have significant impacts on supplies for agricultural, industrial, and domestic uses. Hydroelectric power generation will be affected, for instance with increased production in New Zealand and Scandinavia—at least for some decades—but decreases in the rest of Europe. All of these changes will occur as the world's population grows, bringing an increasing demand for food, energy, and industrial output. As these all depend on mountain water, this increases the imperative for more effective planning and cooperation among all the different stakeholders to ensure that water supplies, when and wherever they are available, are used as effectively as possible.

A further set of challenges will come from the melting of permafrost: soil, bedrock, or any other material that remains below freezing for years. This is found extensively in the alpine and nival zones of mountains, usually below an 'active layer' that thaws and refreezes seasonally. As the climate warms, and snowpacks become shallower and melt earlier, the active layer will become deeper and more extensive. This has happened since the early 20th century in the Alps, and especially from the 1980s, leading to increasing numbers of large rockfalls and rock avalanches, mainly from altitudes above 2,800 metres. About half of these were from locations which had recently been deglaciated. The melting of permafrost has also affected the infrastructure of ski areas, when the permafrost under towers for ski lifts began to melt, and small landslides have occurred. In mountains around the world, such events are likely to become more common and to occur at lower altitudes, affecting not only ski areas, but settlements and transport infrastructure. Another natural hazard that may become more frequent is fire, especially in mountains where summers become hotter and drier. For instance, as summers warmed in the Swiss Alps from the 1980s to the 2000s, the proportion of lightning-caused fires increased from 20 to 41 per cent.

Possible changes in the frequency of fires are one of many challenges for communities and companies that rely on mountain forests for their livelihoods or income. Given the long lifespan of most trees, planning ahead, including deciding which species or varieties to plant, is particularly difficult in the context of climate change. Warmer temperatures, and higher atmospheric concentrations of carbon dioxide, may bring some potentially positive benefits for mountain forests where soils and moisture conditions are suitable: growth rates may increase; tree species may be able to move upwards, thus increasing the diversity of species and habitats; and a higher treeline may improve protection against some natural hazards. However, increases in both tree growth and forest density and cover can also mean that there is more fuel for fires. A further set of challenges relates to outbreaks of pests and diseases: organisms with far shorter lifespans than trees.

Since the mid-1990s, a major outbreak of mountain pine beetle has affected hundreds of thousands of square kilometres of lodgepole pine forest across western North America, killing the majority of the trees either directly or indirectly, because weakened trees are more susceptible to fungi. One set of reasons for the outbreak is climatic: a generation of beetles now only takes one year instead of two, because summers have been hot and dry and winters have been mild—which has also allowed more beetles to survive. A second set is historical: most of the forests are quite uniform, with dense stands of old trees—due to a large extent to past human activities, including both extensive fires in the late 19th century and subsequent fire suppression. This has allowed the outbreak to spread particularly fast. While its scale is unprecedented, the beetles are also moving upwards into whitebark pine forests, and similar events are likely to occur in other mountain regions, affecting both native and imported tree species. For example, Monterey pine, the most common plantation species in New Zealand, may grow faster under climate change; but so might populations of the fungi which cause needle blight, thus offsetting the higher growth. Such processes may have further impacts. Dead and weakened forests may provide more timber for harvesting, but profitable markets for such low-quality timber may not exist, thus threatening the livelihoods of forest-dependent communities. Such forests are more likely to burn, endangering these communities—as well as the animals that live in them—and also meaning that the forests change from carbon stores to carbon sources. The loss of mature forests also results in less effective protection against natural hazards and removes habitats for certain species: it is predicted that boreal owls will be scarce for forty to sixty years after the mountain pine beetle epidemic ends. On the other hand, other species benefit from all these processes—at least in the short term.

While most crops in mountain areas are grown annually, in contrast to the decades that trees live, many of the opportunities and challenges are comparable. As long as soil conditions are suitable and enough water is available, it may be possible to grow crops at higher altitudes: not only grains, but also other staples such as bananas, maize, and plantains in East Africa. Yields may increase and it may become possible to harvest two crops a year, as suggested for the farmers of Chitral, Pakistan, at 1,500 metres. Thus, some farmers may benefit from climate change, particularly where they have retained a large diversity of varieties and are able to plant these as appropriate microhabitats expand. This has already happened in Bolivia; though the farmers also had to find the necessary water. Similarly, those depending on grazing animals may be able to take them to pastures at higher altitudes, opened up by longer snow-free seasons, thus allowing them to enlarge their herds. All of these potential benefits could increase the food security of rural mountain people, of whom a third worldwide—about a quarter of a billion people, particularly in Asia—are at risk of hunger. Unfortunately, many of these opportunities may be counteracted by other factors. Lack of suitable water supplies is one of these, but so is the likelihood that pests and diseases will spread to higher altitudes, so that crop yields decrease and more stored food is lost. Globally, therefore, risks of hunger and malnutrition in mountain areas may well increase unless significant actions are taken. These include benefitting from the existence of crop varieties that are adapted to diverse conditions and the traditional knowledge needed to ensure reliable yields, for both subsistence and sale; greater use of organic farming methods; improving the integration of agriculture, forestry, aquaculture, and local food processing to help diversify income sources and increase the resilience of food systems; and also using wild food sources sustainably—which may require careful negotiation with those responsible for protected areas, especially if these are also seen to be threatened by the upward movement of agriculture.

Malnutrition makes people more susceptible to disease, and a further major concern for mountain people in developing countries is the upwards spread of diseases, especially malaria. The mosquitoes that cause this have moved upwards in recent years as temperatures have warmed: they have reached 2,000 metres in East Africa and 2,200 metres in Bolivia. Malaria has also re-emerged in central China, linked to both warmer temperatures and higher rainfall. Such trends are particularly unwelcome given that mountains in such regions have historically been relatively healthy in comparison to adjacent lowlands. Nevertheless, the means to control malaria are well-known, as long as the necessary resources are available to implement them. Similarly, encephalitis has been reported at higher altitudes in India and Nepal, but this can be treated with vaccines.

Health concerns are also relevant for those hoping to attract tourists. In some regions, especially those relying on snow for skiing, the future of this industry does not look good. This is particularly true for Australia's Snowy Mountains: it has been estimated that maintaining skiing until 2020 will require $100 million of investment, particularly for 700 snow guns which will use 2.5–3.3 gigalitres of water a month. Moreover, the ski areas of New Zealand are not so far away, and snowfall there should remain more reliable; though snow guns are already being used to ensure good skiing conditions (see Figure 20). In the Alps and other mountain regions around the world, low-altitude resorts are unlikely to be able to rely on natural snowfall, and snowmaking will become more unreliable and expensive. However, although summer temperatures are likely to increase in mountains around the world, they will almost always be cooler than urban centres—the main source of tourists—or coasts, the other major environment which is most visited by tourists. Tourism may

20. Snow guns at Whakapapa skifield, Mount Ruapehu, New Zealand. In an era of climate change, snow making can allow ski resorts to improve the reliability of their snow cover and extend their ski seasons—but for how long?

therefore remain an important economic sector for many mountain areas although, as mentioned in Chapter 6, it will be important to ensure that it is integrated well with other sectors. This is also true for *amenity migration*—the movement of people to areas which they perceive as being more attractive to live in, a growing phenomenon in many mountain areas around the world—and other migration from lowland areas into the mountains as lowland climates warm, become less healthy, or because mountains are perceived to offer new opportunities for agricultural and other livelihoods.

Despite the many complex challenges, mountains may also provide particular opportunities with regard to two of the major imperatives deriving from climate change: minimizing emissions of greenhouse gases or storing carbon; and the generation of renewable energy. Mountain forests represent significant stores of carbon, and their potential for being managed to maximize this value is increasingly being recognized in United Nations 'Reducing Emissions from Deforestation and Forest Degradation' (REDD+) projects. One example is a project involving the Asia Network for Sustainable Agriculture and Bioresources, the International Centre for Integrated Mountain Development, and the Federation of Community Forestry Users Nepal, being implemented in 10,266 hectares of forest in Nepal managed by 104 community use groups. Their activities include identifying drivers of deforestation and degradation, measuring carbon stocks, planting fodder species and trees for fuel and slope stabilization, improved forest management, control of grazing, fire prevention, and facilitating the use of biogas and improved cooking stoves. These activities have led to increases in carbon stocks since the project began in 2011, for which the communities receive payment that they can reinvest both in their forests and in other activities to support local livelihoods. In addition to forests, mountain peatlands, grasslands, and shrublands represent important carbon stores, mainly below ground. The management of moorland to minimize carbon emissions is now being explored by a number of land owners in the British uplands. Improved grazing practices, ecosystem restoration, and fire management could also decrease emissions from grasslands and shrublands around the world; this is being explored in the Andes, Australia, Nepal, and New Zealand. There is potential to finance all of these activities through carbon markets and carbon finance mechanisms such as REDD+.

The topography and climate of mountain areas also mean that they have significant potential for the generation of renewable energy. While many sites for large-scale hydroelectricity generation have already been developed, especially in Europe and North America and increasingly in China—and by Chinese companies in other countries—many more sites could be developed. Choices about which sites to develop will, however, be increasingly difficult as they must take into account a wide range of social, environmental, and equity concerns. In addition, there is considerable potential for the installation of much more small or micropower capacity, which can both support local economies and contribute to the mitigation of climate change. This is also true for solar and wind power in mountain areas. While solar power arrays in mountains are generally quite small, typically providing energy for homes and community facilities, wind turbine developments are often large and more intrusive in mountain landscapes. As a consequence, there have been many debates about where they can be sited, with the result that, for instance, the Cairngorms National Park in Scotland is becoming ringed by a series of wind farms, which cannot be located in the park but may be installed where they do not have significant visual impact when seen from the park—a challenging evaluation to make.

## Partnerships for an uncertain future

Climate change emerged on to the global political agenda at the same time as sustainable development. The Rio Earth Summit in 1992 focused on both, resulting in the United Nations Framework

Convention on Climate Change and 'Agenda 21', both of which specifically consider mountains. Sustainable mountain development, a term introduced in the mountain chapter of 'Agenda 21', refers to the continued well-being, first, of mountain people and the environments on which they rely and, second, of the billions of people who depend on the many goods and services that mountains provide—even though this may not be recognized by most of them, or their governments. Ensuring this well-being in the long term represents very significant challenges in an era characterized by globalization, economic challenges, and uncertain outlooks for supplies of food, energy, and many other resources. Increasing attention to sustainable mountain development therefore requires constructive and informed cooperation between the many stakeholders concerned with mountain areas, whether they live in these areas; in capital cities, which are often far from the mountains; or in other countries that provide financial and other resources, often through international organizations. In this context, the challenges of climate change principally provide an additional impetus towards cooperation, whether this is in scientific research, training and education, sharing of knowledge and data, policy development and implementation, or in equitable transfers of financial resources. A number of structures for cooperation in such fields have emerged in recent years, such as the Mountain Research Initiative, the Mountain Partnership, the Mountain Initiative for Climate Change initiated by the government of Nepal in 2009, National Adaptation Programmes for Action on climate change in many developing countries, and the project on ecosystem-based adaptation in mountain areas funded by the German government and implemented by IUCN, UNEP, and UNDP—initially in Nepal, Peru, and Uganda.

Although many mountain people are known for being physically tough and often militaristic—and have fought and provided mercenaries in many wars—another common characteristic of most traditional mountain societies is that they have had to cooperate in order to ensure their long-term survival in environments where, often, growing seasons are short, slopes are steep, soils are thin, and water supplies must be managed carefully. As a result, these societies have carefully crafted institutions to manage their forests, grazing land, irrigation systems, and other resources. Such institutions still exist in many mountain areas, though many have also been weakened or lost as these societies have been increasingly influenced by external forces such as emigration for paid employment, changing economic situations, and the actions of governments. Nevertheless, these institutions show how people with often divergent interests can come together to act towards the common good. They may also provide valuable models for new partnerships for mountain areas in a globalized world. These may involve very different stakeholders, including national and international NGOs, private companies—such as trekking companies which also undertake community development projects—and research and development organizations, as with the REDD+ project in Nepal mentioned in the section entitled 'Challenges and opportunities'. While partnerships of national and international scales are also emerging, more are needed to ensure that, even in this unpredictable era of climate change, mountain people and environments can have a sustainable future—to the benefit of all of us.

# Further reading

The books and websites listed below include some that cover themes addressed in a number of chapters; others are more relevant to a single chapter. Most have a global, continental, or regional focus; there are very many others on specific mountain ranges or even villages, which are not included here. The best place to look for books on mountain topics is on the website of the journal *Mountain Research and Development* and the reviews in that journal.

## General sources on mountains

C. Ariza, D. Maselli, and T. Kohler, *Mountains: Our Life, our Future. Progress and Perspectives on Sustainable Mountain Development* (Bern: Swiss Agency for Development and Cooperation and Centre for Development and Environment, 2013).

European Environment Agency, *Europe's Ecological Backbone: Recognising the True Value of our Mountains* (Copenhagen: European Environment Agency, 2010).

B. Messerli and J.D. Ives (eds.), *Mountains of the World: A Global Priority* (New York and London: Parthenon, 1997).

M.F. Price (ed.), *Mountain Area Research and Management: Integrated Approaches* (London: Earthscan, 2007).

M.F. Price, L. Jansky, and A.A. Iatsenia (eds.), *Key Issues for Mountain Areas* (Tokyo: United Nations University Press, 2004).

M.F. Price, A.C. Byers, D.A. Friend, T. Kohler, and L.W. Price (eds.), *Mountain Geography: Physical and Human Dimensions* (Berkeley: University of California Press, 2013).

D.B.A. Thompson, M.F. Price, and C.A. Galbraith (eds.), *Mountains of Northern Europe: Conservation, Management, People and Nature* (Edinburgh: The Stationery Office, 2005).

Alpine Convention: <http://www.alpconv.org/pages/default.aspx>.

CIPRA: <http://cipra.org/en>: the best starting point for current information on the Alps.

International Centre for Integrated Mountain Development: <http://www.icimod.org/>: an essential site for information on the Hindu Kush-Himalaya.

Mountain Forum: <http://www.mtnforum.org>: a knowledge repository, social network, and information portal for people interested in sustainable mountain development.

Mountain Research and Development: <http://www.mrd-journal.org>: the longest-established global mountain journal, including articles, papers, book reviews, and other material. All issues from 2000 are open-access.

*Revue de Geographie Alpine / Journal of Mountain Research*: <http://rga.revues.org/index.html>: a valuable journal published since 1913. All issues are available online.

## Chapter 1: Why do mountains matter?

R.G. Barry, *Mountain Weather and Climate* (Cambridge: Cambridge University Press, 2008).

E. Bernbaum, *Sacred Mountains of the World* (San Francisco: Sierra Club Books, 1997).

B. Debarbieux and G. Rudaz, *The Mountain Makers* (Chicago: University of Chicago Press, 2015).

J.D. Ives, *Sustainable Mountain Development—Getting the Facts Right* (Lalitpur: Himalayan Association for the Advancement of Science, 2013).

J. Mathieu, *The Third Dimension: A Comparative History of Mountains in the Modern Era* (Cambridge: White Horse Press, 2011).

National Research Council, *Lost Crops of the Incas* (Washington, DC: National Academy Press, 1989).

## Chapter 2: Mountains are not eternal

G. Heiken, *Dangerous Neighbours: Volcanoes and Cities* (Cambridge: Cambridge University Press, 2013).

M.R.W. Johnson and S.L. Harley, *Orogenesis—The Making of Mountains* (Cambridge: Cambridge University Press, 2012).

P. Owens and O. Slaymaker (eds.), *Mountain Geomorphology* (London: Edward Arnold, 2004).

## Chapter 3: The world's water towers

U. Bundi (ed.), *Alpine Waters* (Heidelberg: Springer, 2010).

T. Hofer and B. Messerli, *Floods in Bangladesh: History, Dynamics and Rethinking the Role of the Himalayas* (Tokyo: United Nations University Press, 2006).

Ellen E. Wohl (ed.), *Inland Flood Hazards: Human, Riparian, and Aquatic communities* (Cambridge: Cambridge University Press, 2000).

## Chapter 4: Living in a vertical world

L.L. Bruijnzeel, F.A. Scatena, and L.S. Hamilton (eds.), *Tropical Montane Cloud Forests: Science for Conservation and Management* (Cambridge: Cambridge University Press, 2010).

I. Coxhead and G.E. Shively (eds.), *Land Use Changes in Tropical Watersheds: Evidence, Causes and Remedies* (Wallingford: CABI, 2005).

H.L. Fröhlich, P. Schreinemachers, K. Stahr, and G. Clemens (eds.), *Sustainable Land Use and Rural Development in Southeast Asia: Innovations and Policies for Mountainous Areas* (Heidelberg: Springer, 2013).

L. German, J. Mowo, T. Amede, and K. Masuki (eds.), *Integrated Natural Resource Management in the Highlands of Eastern Africa—From Concept to Practice* (Abingdon: Earthscan, 2012).

C. Körner, *Alpine Treelines: Functional Ecology of the Global High Elevation Tree Limits* (Dordrecht: Springer, 2012).

H-P. Liniger and W. Critchley (eds.), *Where the Land is Greener—Case Studies and Analysis of Soil and Water Conservation Initiatives Worldwide* (Wageningen: CTA, UNEP, FAO, and CDE, 2007).

M.F. Price, G. Gratzer, L.A., Duguma, T. Kohler, D. Maselli, and R. Romeo (eds.), *Mountain Forests in a Changing World—Realizing Values, Addressing Challenges* (Rome: Food and Agriculture Organization of the United Nations, 2011).

M.K. Steinberg, J.J. Hobbs, and K. Mathewson (eds.), *Dangerous Harvest: Drug Plants and the Transformation of Indigenous Landscapes* (Oxford: Oxford University Press, 2004).

## Chapter 5: Centres of diversity

B. Brower and B.R. Johnston (eds.), *Disappearing People? Indigenous Groups and Ethnic Minorities in South and Central Asia* (Oxford: Berg/Left Coast Press, 2007).

C. Körner and E.M. Spehn (eds.), *Mountain Biodiversity—A Global Assessment* (New York and London: Parthenon, 2002).

S.A. Laird, R. McLain, and R.P. Wynberg (eds.), *Wild Product Governance: Finding Policies That Work for Non-timber Forest Products* (London: Earthscan, 2010).

L. Nagy and G. Grabherr, *The Biology of Alpine Habitats* (Oxford: Oxford University Press, 2009).

R.E. Rhoades (ed.), *Development with Identity: Community, Culture and Sustainability in the Andes* (Wallingford: CABI Publishing, 2006).

E.M. Spehn, M. Liberman, and C. Körner (eds.), *Land Use Change and Mountain Biodiversity* (Boca Raton: CRC Press, 2006).

S. Stevens (ed.), *Indigenous People, National Parks, and Protected Areas* (Tucson: University of Arizona Press, 2014).

## Chapter 6: Protected areas and tourism

Austrian MAB Committee, *Biosphere Reserves in the Mountains of the World: Excellence in the Clouds?* (Vienna: Austrian Academy of Sciences Press, 2011).

B. Debarbieux, M. Oiry Varacca, G. Rudaz, D. Maselli, T. Kohler, and M. Jurek (eds.), *Tourism in Mountain Regions: Hopes, Fears and Realities* (Geneva: University of Geneva, 2014).

D. Harmon and G.L. Worboys (eds.), *Managing Mountain Protected Areas: Challenges and Responses for the 21st Century* (Colledara: Andromeda, 2004).

B. Verschuuren, R. Wild, J.A. McNeely, and G. Oviedo (eds.), *Sacred Natural Sites: Conserving Nature and Culture* (London: Earthscan, 2010).

G. Worboys, W.L. Francis, and M. Lockwood (eds.), *Connectivity Conservation Management—A Global Guide* (London: Earthscan, 2010).

Mountain Protected Areas Network: <http://conservationconnectivity. org/mountains-wcpa/about.htm>.

Mountain Voices: <http://mountainvoices.org/>: a unique set of oral testimonies from ten mountain regions.

## Chapter 7: Climate change in the mountains

A. Bonn, T. Allott, K. Hubacek, and J. Stewart (eds.), *Drivers of Environmental Change in Uplands* (Abingdon: Routledge, 2009).

U.M. Huber, H.K.M. Bugmann, and M.A. Reasoner (eds.), *Global Change and Mountain Regions—An Overview of Current Knowledge* (Dordrecht: Springer, 2005).

T. Kohler, A. Wehrli, and M. Jurek (eds.), *Mountains and Climate Change: A Global Concern* (Bern: Centre for Environment and Development, 2014).

L.A.G. Moss and R.S. Glorioso (eds.), *Global Amenity Migration—Transforming Rural Culture, Economy and Landscape* (Kaslo and Port Townsend: New Ecology Press, 2014).

B. Orlove, E. Wiegandt, and B.H. Luckman (eds.), *Darkening Peaks: Glacier Retreat, Science and Society* (Berkeley: University of California Press, 2008).

Mountain Partnership: <http://www.mountainpartnership.org>: a valuable source of information on organizations working on mountain issues, with many key documents.

Mountain Research Initiative: <http://mri.scnatweb.ch/en/>: a global initiative supporting research on all aspects of global change in the world's mountains.

# Index

Mountains